36 Advances in Polymer Science

Fortschritte der Hochpolymeren-Forschung

Edited by H.-J. CANTOW, Freiburg i. Br. · G. DALL'ASTA, Colleferro
K. DUŠEK, Prague · J. D. FERRY, Madison · H. FUJITA, Osaka
M. GORDON, Colchester · W. KERN, Mainz · S. OKAMURA, Kyoto
C. G. OVERBERGER, Ann Arbor · T. SAEGUSA, Kyoto · G. V. SCHULZ, Mainz
W. P. SLICHTER, Murray Hill · J. K. STILLE, Fort Collins

With 30 Figures

Springer-Verlag
Berlin Heidelberg GmbH 1980

Editors

Prof. Dr. HANS-JOACHIM CANTOW, Institut für Makromolekulare Chemie der Universität, Stefan-Meier-Str. 31, 7800 Freiburg i. Br., BRD

Prof. Dr. GINO DALL'ASTA, SNIA VISCOSA – Centro Studi Chimico, Colleferro (Roma), Italia

Prof. Dr. KAREL DUŠEK, Institute of Macromolecular Chemistry, Czechoslovak Academy of Sciences, 162 06 Prague 616, ČSSR

Prof. JOHN D. FERRY, Department of Chemistry, The University of Wisconsin, Madison, Wisconsin 53706, U.S.A.

Prof. HIROSHI FUJITA, Department of Polymer Science, Osaka University, Toyonaka, Osaka, Japan

Prof. MANFRED GORDON, Department of Chemistry, University of Essex, Wivenhoe Park, Colchester C04 3 SQ, England

Prof. Dr. WERNER KERN, Institut für Organische Chemie der Universität, 6500 Mainz, BRD

Prof. SEIZO OKAMURA, No. 24, Minami-Goshomachi, Okazaki, Sakyo-Ku, Kyoto 606, Japan

Prof. CHARLES G. OVERBERGER, Macromolecular Research Center, Institute of Science and Technology, The University of Michigan, Ann Arbor, Michigan 48 104, U.S.A.

Prof. TAKEO SAEGUSA, Department of Synthetic Chemistry, Faculty of Engineering, Kyoto University, Kyoto, Japan

Prof. Dr. GÜNTER VICTOR SCHULZ, Institut für Physikalische Chemie der Universität, 6500 Mainz, BRD

Dr. WILLIAM P. SLICHTER, Chemical Physics Research Department, Bell Telephone Laboratories, Murray Hill, New Jersey 07971, U.S.A.

Prof. JOHN K. STILLE, Department of Chemistry, Colorado State University, Fort Collins, Colorado 805 23, U.S.A.

ISBN 978-3-662-15390-1   ISBN 978-3-540-38275-1 (eBook)
DOI 10.1007/978-3-540-38275-1

Library of Congress Catalog Card Number 61-642

© by Springer-Verlag Berlin Heidelberg 1980

Originally published by Springer-Verlag Berlin Heidelberg New York in 1980.

Softcover reprint of the hardcover 1st edition 1980

2152/3140 – 543210

Contents

Unperturbed Dimensions of Stereoregular Polymers

Robert Jenkins and Roger S. Porter

Materials Research Laboratory, Polymer Science and Engineering Department
University of Massachusetts, Amherst, Massachusetts 01003, U.S.A.

Results from numerous studies of unperturbed dimensions for vinyl polymers are tabulated for the first time. A comparison is made between results from theta temperature measurement and those estimated from measurements in good solvents. Data for poly(1-pentene)[38] was used by the authors to obtain Figs. 2–5. This data had not previously been used to obtain estimates of the uperturbed dimensions. It is concluded that these estimation procedures are generally quite acceptable, giving values usually within 8% of those determined in theta solvents. Statistical calculations for the dependence of the unperturbed dimensions on stereoregularity are considered in the light of the available data. The calculations predict an increase in the characteristic ratio, C_∞, at intermediate tacticities for vinylidene polymers whereas a decrease is predicted for vinyl polymers. It is shown that little experimental data currently exists to confirm these predictions. It is pointed out that this is not a result of insufficient experimental studies, but rather insufficient characterization of those polymers which were studied.

Table of Contents

I. Introduction

Since the initial advancements in the synthesis of stereoregular polymers by Natta et al.[1] in the late fifties, attempts have been made to obtain stereoregular forms of many polymers. With the synthesis of these new polymers came a large number of reports on their properties, both in the bulk state and in solution. Among the properties studies have been crystallinity, and the effect of stereoregularity on the unperturbed dimensions. In this review we are concerned exclusively with the latter. A thorough understanding of the effect of stereoregularity on the unperturbed dimensions is essential since they are a direct result of the chain conformations. Knowledge of the effect of stereoregularity on a chains conformations is important since the conformations determine many of the properties of polymers, both in the bulk state and in solution. Properties such as rubber elasticity, the hydrodynamics and thermodynamics of polymer solutions, and optical properties are only several of the properties that are dependent on chain conformations. Although it is now possible to obtain specific stereoregular forms for many polymers, we are concerned here only with stereoregular vinyl and vinylidene polymers.

There appears to be no prior reviews concerned with the effect of stereoregularity on unperturbed chain dimensions. Our intent is to compile the reliable data and to elucidate whatever trends may exist. This review also reveals areas that suggest further study. Admittedly, the data given in many instances is questionable. One of the many obstacles in any such study is the exactness with which stereoregularity has been determined. In all early publications little could be reported about the absolute stereoregularity. Commonly researchers have referred to polymers as "isotactic", "syndiotactic", or simply "atactic", with little reference as to the degree. Still others chose to refer to polymer products by the method with which they were synthesized, or by the fraction number, as collected from polymer fractionation procedures. Today, with the advancements made in both polymer synthesis and characterization techniques, especially nuclear magnetic resonance spectroscopy, (NMR), an increasing body of knowledge is becoming available from studies of polymers of known stereoregularity. However, data from systematic studies of well characterized stereoregular polymers is still scarce, and the data reported in this review should be received with this in mind.

II. Historical

The first systematic evaluation of chain dimensions of stereoregular polymers in solution were reported by Danusso and Moraglio[2]. They studied both isotactic and atactic polystyrene by viscometry and osmometry, in benzene and in toluene, (both thermodynamically good solvents). They concluded that no difference could be observed in the intrinsic viscosity-molecular weight relationship, $[\eta] = k'M^a$, for the two stereoregular forms [see Fig. 1]. However, they did find a noticeable difference between the second varial coefficients for the two stereoregular forms of equivalent molecular weight. These results were consistent with subsequent measurements on

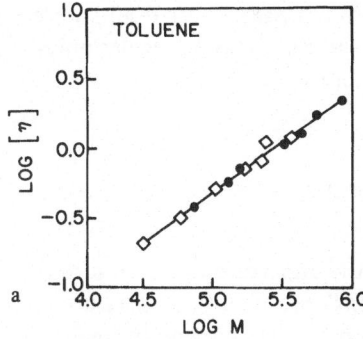
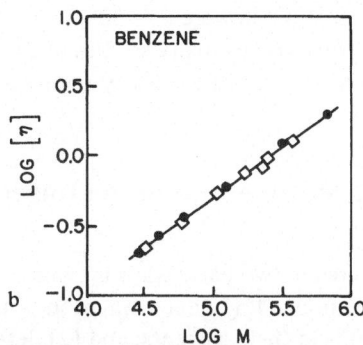

Fig. 1. a Relation between [η] in toluene and M for fractions of polystyrene; (◇) isotactic polystyrene; (●) atactic polystyrene. From Ref.[2]. b Relation between [η] in benzene and M for fractions of polystyrene; (◇) isotactic polystyrene; (●) atactic polystyrene. From Ref.[2]

isotactic[3–8] and atactic[7, 9–12] polystyrene (PS);
isotactic[13–18], atactic[13–16], and syndiotactic[16] polypropylene (PP);
isotactic and atactic poly(1-pentene)[19] (PPT);
isotactic[10, 20], atactic[21, 22], and syndiotactic[23] poly(methyl methacrylate) (PMMA);
and isotactic, atactic, and syndiotactic poly(isopropyl acrylate)[24, 25] (PIPA);

All of these systems showed that, in thermodynamically good solvents, no differences are measurable in the intrinsic viscosity-molecular weight relationship for the different stereoregular forms. In addition, the second varial coefficient, A_2, in every case studied was found to be larger for the syndiotactic and atactic forms than for the isotactic forms[7, 8, 14, 15, 19, 25, 26].

The chain dimensions of polymer molecules in solution are influenced by both long range (excluded volume), and short range (rotational isomeric) effects[27]. Long range effects are a result of thermodynamic interactions between polymer molecules and their environments. In a good solvent, where the energy of interaction between polymer and solvent is high, the molecules will tend to expand in order to increase the number of polymer-solvent contacts. As a result, the volume which one polymer segment excludes from another is large. In a poor solvent where the energy of interaction is unfavorable, the polymer will contract in order to increase the amount of polymer-polymer contacts. This decreases the amount of polymer-solvent contacts, resulting in a low excluded volume. Under theta conditions, defined as the temperature at which the second varial coefficient is zero[27], it has been shown that the excluded volume vanishes and the chain is unperturbed by long range interactions. Here the dimensions are simply a function of the short range effects and are therefore referred to as the unperturbed dimensions. Short range effects are a result of the rotational isomeric states available to the particular molecule, and hence are indicative of the conformations of the polymer molecule itself.

Krigbaum et al.[8], were the first to emphasize the necessity of measuring the dimensions of stereoregular polymers unperturbed by long range effects. They used light scattering and intrinsic viscosity measurements in thermodynamically good solvents to obtain the unperturbed dimensions of isotactic PS. Comparison with the value for atactic PS[9] led to the conclusion that the unperturbed dimensions of the isotactic form were 25–30% larger than the atactic counterpart. These results were the first among many studies of the unperturbed dimensions of stereoregular polymers, and served to clearly point the way for subsequent work.

III. Measurement of the Unperturbed Dimensions

There are two basic ways in which measurements of the unperturbed dimensions are obtained: (1) determination of unperturbed dimensions directly, by measurements in theta solvents; and (2) determination of the perturbed dimensions in a good solvent and extrapolation of the values to the unperturbed state using one of the existing theories. Both methods have been widely used as will be shown.

A. Direct Measures of Unperturbed Dimensions

1. Methods

The unperturbed dimensions can be directly determined by methods such as light scattering and dilute solution viscometry on polymers dissolved in theta, θ, solvents. The light scattering technique involves graphical methods as outlined by Zimm[28] which involves obtaining light scattering data from a number of solute concentrations, each at several different scattering angles. Eq. 1 is used:

$$\frac{Kc}{R_\theta} = \frac{1}{M_w}(1 + (16\,\pi^2/3\,\lambda^{1/2})\sin^2(\theta/2)\,\langle\overline{S}\rangle_z^2) + 2\,A_2 C + \ldots \tag{1}$$

where; $K = 2\,\pi N_0^2 (dn/dc)^2/N_a \lambda^4$, c is the solute concentration, R_θ is the Raleigh ratio, and θ is the scattering angle. The ratio Kc/R_θ is plotted as a function of $\sin^2(\theta/2) + qc$, where q is an arbitrary constant. The data obtained at a given scattering angle are then extrapolated to $c = 0$. Then the data obtained at different scattering angles for the same concentrations are extrapolated to $\theta = 0$. Through this double extrapolation procedure, it is possible to eliminate effects due to deviations from solution ideality as well as effects due to destructive interference of the scattered light. It follows from Eq. 1, that the mean-square-radius of gyration, $\langle\overline{S}\rangle_z^2$, may be calculated from the initial slope of the $c = 0$ line from the Zimm plot. The radius of gyration from such studies is the z-average, and must be corrected for heterogeneity to obtain $\langle\overline{S}\rangle_0^2$. Assuming a linear polymer, the mean square end-to-end distance, $\langle\overline{r}_0\rangle^2$ is:

$$\langle\overline{r}_0\rangle^2 = 6\langle\overline{S}_0\rangle^2 \tag{2}$$

Although theoretically sound, this method has proved to be difficult to impossible, in practice. In general, the light scattering method is hindered by a number of errors: optical artifacts, molecular heterogeneity, and the errors inherent in the dual extrapolation required with respect to concentration and angle. In addition, experimental difficulties, such as crystallization and fractional precipitation, often encountered when working with stereoregular polymers near theta conditions are generally formidable. As a result there are relatively few reports of light scattering measurements of stereoregular polymers in theta solvents[29, 30, 30a].

Viscosity measurements on polymers dissolved in theta solvents have proved more useful. However, again measurements are subject to difficulties due to poor solubility near the theta point. At the theta temperature the Flory-Fox relationship[31] reduces to:

$$[\eta]_\theta = K_\theta M^{1/2} \tag{3}$$

where:

$$K_\theta = \Phi (6 \langle \overline{S}_0 \rangle^2 /\overline{M})^{3/2} = \Phi (\langle \overline{r}_0 \rangle^2 /M)^{3/2} \tag{4}$$

and: Φ is the "universal constant" of Flory, and $\overline{M}$ is the number average molecular weight[27]. Consequently measurements of the intrinsic viscosity at the θ temperature for samples of known molecular weights allows calculation of K_θ. Determination of the constant Φ thus allows calculation of the unperturbed parameter $(\langle \overline{r}_0 \rangle^2 /M)^{1/2}$, as well as the characteristic ratio, C_∞, defined in Eq. 5:

$$C_\infty = (K_\theta / \Phi)^{2/3} M_b /\ell^2 \tag{5}$$

where M_b is the mean molecular weight per skeletal bond, ℓ is the bond length, and Φ is again the Flory constant.

Flory[27] has stated that Φ should be independent of molecular characteristics, and should therefore be the same for all randomly coiled chains. According to Flory[27], the best value for Φ is $2.1 (\pm 0.2) \times 10^{21}$, which, when corrected for heterogeneity, should yield a value of 2.5×10^{21}. This value is substantiated by Cowie[32], who studied anionically-produced polystyrene in several solvents, and found an average of 2.55×10^{21}. However, many authors have determined Φ from light scattering and dilute solution viscometry, and have suggested different values. Ptitsyn et al.[33], as well as Stockmayer and Kurata[34], have noted that Φ may vary from 1.9×10^{21} for measurements in good solvents to 2.9×10^{21} for measurements in theta solvents. In addition, Mark et al.[25] determined Φ for isotactic, atactic and syndiotactic poly(isopropyl acrylate) and found values of $2.0 (\pm 0.2) \times 10^{21}$, $2.5 (\pm 0.2) \times 10^{21}$, and $2.6 (\pm 0.2) \times 10^{21}$ respectively. It appears that Φ is not as constant as originally believed. In addition, the value of Φ used in Eq. 4 for calculation of K_θ as well as C_∞, varies from one author to another. It should also be noted that few researchers have actually determined Φ for their systems. As a result, until more accurate values of Φ are obtained for narrow distribution samples for each stereoregular polymer, in the appropriate solvent, and at specific temperatures, the values reported in Tables 1–7 must be taken as approximations.

2. Stereoregular Systems

The unperturbed dimensions for a number of stereoregular polymers have been studied by dilute solution viscometry in theta solvents. Given in Tables (1–8) are values of K_θ, as defined in Eq. 4, and the corresponding values of C_∞. As can be seen from these eight tables, comparisons can be made between the stereoregular forms for only four polymers; PMMA, PMS, PPT, and PP. For the other three polymers sufficient data from direct, theta point studies is not available. Of the four polymers for which comparisons can be made, we see that the variation in chain dimensions with tacticity is not uniform. For PMMA it appears that the isotactic form has the larger unperturbed dimensions whereas the atactic and syndiotactic forms are equivalent. This latter result may be expected since atactic PMMA is predominantly syndiotactic[35]. For PMS, the isotactic form again exhibits larger unperturbed dimensions than the syndiotactic form. This same result is also found for PPT. However, the reverse trend is observed for PP where the isotactic form has the smaller unperturbed dimensions.

Interpretation of these results is complicated by the fact that the unperturbed parameter, K_θ, used to obtain both $(\langle \overline{r}_0 \rangle^2/M)^{1/2}$ and C_∞, is not strictly a constant. It has been demonstrated in several studies that K_θ is subject to the combined effects of both solvent[32, 36, 37, 38a] and temperature[36, 38, 38a], although the effect of solvent is believed to be small[27]. In principle however, solvation can perturb the relative energies of the conformations accessible to polymer chain segments[39]. Since the unperturbed dimensions are determined by the conformations of the chain, solvent effects may be observed. In general, the unperturbed dimensions may be expected to be susceptible to solvent effects if the chain backbone contains polar groups. This has been shown for both poly(hexene 1-sulfone)[40, 41] and poly(dimethyl siloxane)[42], both of which contain polar backbone bonds. However, measurable solvent effects are also observed[37] for less polar chains such as PS[32, 36, 37], PMMA[37] and poly(vinyl acetate)[37], PVAc. An obvious way of determining the effect of solvent is to compare the unperturbed dimensions of the same polymer in solvents that exhibit the same theta temperature. Consider the data for atactic PS given in Table 2; in 1-chloro-n decane, diethyl malonate, and cyclohexane. All have relatively close theta temperatures (i.e. 39.9, 34.8, 32.8). The differences in the unperturbed parameter K_θ, (7.9, 7.7, + 8.2 x 10^{-4}), must therefore result from solvent effects[36].

The effect of temperature on unperturbed dimensions although straightforward in principle, has proved to be difficult to evaluate experimentally. Based on the rotational isomeric model, one would expect a negative temperature coefficient since as the thermal energy increases the relative populations of the g^+ and g^- states increases over the trans and consequently $(\overline{r}_0^2)^{1/2}$ and therefore K_θ should decrease. However, evaluation of the temperature coefficients for stereoregular polymers have resulted in inconsistencies[43, 44, 45]. These inconsistencies have been attributed in part to different techniques used to obtain the temperature coefficient, d ln $\langle \overline{r}_0 \rangle^2$/dt.

There are two basic ways by which d ln $\langle \overline{r}_0 \rangle^2$/dt may be determined. The first involves measurements in dilute solution[43, 27]; the second, solid state stress-temperature measurements[46, 47]. The dilute solution measurements have been shown to be fraught with problems arising from solvent effects, as discussed in the preceding Sect. However, data obtained from stress-temperature measurements for amorphous networks are believed to be more reliable. The data in Table 9 shows the discouraging pattern of the available measurements of d ln $\langle \overline{r}_0 \rangle^2$/dt for stereoregular PMMA. All data come from solution measurements. Consequently specific effects of solvent

Table 1. Poly(methyl methacrylate)

Tacticity	Solvent	Exp. temp. °C	θ temp. °C	$K_\theta \times 10^4$ Direct determination	(F-F)	(S-F)	(K-S)	(B)	C_∞	$\Phi \times 10^{-21}$	Ref.
Atactic	4-Heptanone	33	33	4.7	–	–	–	–	7.8	2.1[a]	68)
	Acetonitrile	45	45	4.9	–	–	–	–	8.1	2.1[a]	68)
	3-Octanone	72	72	5.0	–	–	–	–	8.2	2.1[a]	68)
	p-Xylene	50	50	4.9	–	–	–	–	7.1	2.5	69)
		60–70	–	–	4.6	4.6	–	–	6.8	2.5	69)
	m-Xylene	30	30	4.9	–	–	–	–	7.1	2.5	69)
		40–70	–	–	4.1	4.4	–	–	6.5	2.5	69)
	o-Xylene	40–70	–3	–	4.1	4.4	–	–	6.5	2.5	69)
	n-Butyl Bromide	35	35	4.6	–	–	–	–	6.8	2.5	69)
		42–58	–	–	4.4	4.6	–	4.6	6.8	2.5	69)
	Isoamyl Acetate	50	50	4.5	–	–	–	–	6.7	2.5	69)
		65–80	–	–	2.4–2.7	3.5–3.7	–	–	5.1	2.5	69)
	Cyclohexane	40–70	–10	–	4.6	7.0	–	4.6	7.6	2.5	69)
	1/1 MEK/ Isoprop.	25	25	5.9	–	–	–	–	9.7	1.9[a]	29,70)
	Chloroform	30	–	–	4.9	–	–	–	7.1	2.5	71)
	Benzene	30	–	–	4.9	–	–	–	7.1	2.5	71)
	MEK	30	–	–	4.9	–	–	–	7.1	2.5	71)
	Acetone	30	–	–	4.9	–	–	–	7.1	2.5	71)
	TFP	25	–	–	–	11.6	11.6	–	11.6	2.87[a]	72)
	p-Cumene	159.7	159.7	5.7	–	–	–	–	7.2	2.87[a]	70)
	n-Propanol	84.4	84.4	6.8	–	–	–	–	8.1	2.87[a]	70)
	3-Heptanone	36.7	36.7	6.3	–	–	–	–	7.7	2.87[a]	70)
Syndio-tactic	n-Butyl Bromide	35	35	5.1	–	–	–	–	7.3	2.5	73)
	Butanone Isopropanol	8	8	4.4	–	–	–	–	6.6	2.5	73)
	2-Heptanone	5–50	30.2	–	4.8	4.8	4.8	4.8	6.5	2.83[a]	74)
	Nitromethane	25–45	15	–	4.8	4.8	4.8	4.8	6.5	2.83[a]	74)
Isotactic	Acetonitrile	27.5	27.5	7.6	–	–	–	–	9.5	2.55[a]	75)
	Acetonitrile	27.6	27.6	7.6	–	–	–	–	10.8	2.1[a]	76)
	n-Butyl Chloride	26.5	26.5	7.7	–	–	–	–	9.5	2.58[a]	73)
	Butanone Isopropanol	25	25	7.2	–	–	–	–	9.1	2.58[a]	73)
	Ethyl Acetate	25	–	–	–	7.7	7.8	7.3	9.4	2.55[a]	75)
	Acetone	30	–	–	7.0	–	–	–	10.1	2.1[a]	20)
	TFP	25	–	–	–	8.2	8.7	–	9.3	2.87[a]	72)
	MEK Isopropanol	30.3	30.3	9.0	–	–	–	–	9.7	2.87[a]	70)
	3-Heptanone	40.0	40.0	8.7	–	–	–	–	9.5	2.87	70)
	n-Propanol	75.9	75.9	7.6	–	–	–	–	8.7	2.87[a]	70)
	p-Cymene	152.1	152.1	5.7	–	–	–	–	7.2	2.87[a]	70)

[a] Indicates the value of Φ used by the respective authors

Table 2. Polystyrene

Tacticity	Solvent	Exp. temp.	θ temp.	$K_\theta \times 10^4$ Direct determination	(F-F)	(S-F)	(K-S)	(B)	C_∞	$\Phi \times 10^{-21}$	Ref.
Atactic	1-Chloro-n-decane	6.6	6.6	7.8	–	–	–	–	10.1	2.5	36)
	1-Chloro-undecane	32.8	32.8	7.9	–	–	–	–	10.2	2.5	36)
	1-Chloro-n-dodecane	58.6	58.6	8.1	–	–	–	–	10.3	2.5	36)
	73%-trans decalin	18	18	7.7	–	–	–	–	10.0	2.5	77)
	100%-trans decalin	24	24	8.2	–	–	–	–	10.4	2.5	77)
	Diethyl malonate	35.9	35.9	7.7	–	–	–	–	10.0	2.5	36)
	Toluene/ n-heptane	30	30	8.6	–	–	–	–	10.8	2.5	27)
	Cyclohexane	34.8	34.8	8.2	–	–	–	–	10.4	2.5	78)
	Methyl cyclo-hexane	68	68	7.8	–	–	–	–	10.1	2.5	77)
	Cyclohexane	34	34	8.0	–	–	–	–	10.3	2.5	79)
	Ethyl cyclo-hexane	70	70	7.3	–	–	–	–	9.6	2.5	79)
Isotactic	PCT	25		10.0^b	–	–	–	–	12.2	2.4^a	8)
	Benzene	30			–	–	9.0	–	11.1	2.5	6)

a   Indicates the value of Φ used by the respective authors. Where no value of Φ was suggested, we chose to
    use 2.5×10^{21}
b   Obtained by A_2 data by the method of Flory and Orifino[56]

are not eliminated. The more reliable method of determination of $d \ln \langle \overline{r}_0 \rangle^2 / dt$ from
the temperature coefficient of stress exhibited by a strained network has not been
applied to PMMA. In general, quantitative comparisons of unperturbed dimensions
for the stereoregular polymers, (see Tables 1—8), may be made only after both the
temperature and solvent effects have been evaluated.

B. Indirect Measures of Unperturbed Dimensions

1. Methods

The attainment of θ conditions, as previously discussed, is often experimentally
difficult. This is especially true for certain stereoregular polymers which have a
tendency to crystallize as well as to fractionally precipitate out from poor solvents.
It is therefore necessary to have a method whereby the unperturbed dimensions
may be obtained from measurements in thermodynamically good solvents. Several

Table 3. Poly(α-methyl styrene)

Tacticity	Solvent	Exp. temp. °C	θ temp. °C	$K_\theta \times 10^4$ Direct determi- nation	(F-F)	(S-F)	(K-S)	(B)	C_∞	$\Phi \times 10^{-21}$	Ref.
Syndio- tactic	trans decalin	9.5	9.5	6.7	–	–	–	–	10.3	2.5	80)
	t-Decalin/ cyclohexane 1/1	17	17	6.8		–	–	–	10.4	2.5	80)
	Cyclohexane	34.5	34.5	7.3	–	–	–	–	10.9	2.5	80)
	t Dec./methyl cyclohexane 2/5	55.5	55.5	6.9	–	–	–	–	10.5	2.5	80)
	Cyclohexane/ methyl cyclohexane 1/1	58.6	58.6	7.4	–	–	–	–	11.0	2.5	80)
	t-Dec./methyl cyclohexane 1/10	81	81	6.8	–	–	–	–	10.4	2.5	80)
↓	Cyclohexane	32.5	32.5	6.6	–	–	–	–	10.2	2.5	81)
Isotactic	Cyclohexane	37	37	7.8	–	–	–	–	11.4	2.5	81)
	Toluene	30	–	–	–	7.5	–	–	11.1	2.5	48)
↓	Methanol benzene 20/80	30	30	7.68	–	–	–	–	11.3	2.5	82)

Note: Characterization of stereoregular PMS is a subject of considerable controversy. Many disagree as to whether a specific sample is syndiotactic or isotactic. As a result, the method of polymerization is usually referred to in the literature instead of the stereoregularity.

Table 4. Poly(isopropyl acrylate)

Tacticity	Solvent	Exp. temp. °C	θ temp. °C	$K_\theta \times 10^4$ Direct determi- nation	(F-F)	(S-F)	(K-S)	(B)	C_∞	$\Phi \times 10^{-21}$	Ref.
Atactic	Benzene	25	–	–	–	–	–	–	7.1[b]	2.5±2[a]	25)
↓	TFP	25	–	–	9.2	–	13.7	–	14.3	2.5	24)
Syndio- tactic	Bromobenzene	60	–	–	–	–	–	–	7.2[b]	2.6±2[a]	25)
Isotactic	Bromobenzene	60	–	–	–	–	–	–	9.7[b]	2.0±2[a]	25)
↓	TFP	25	–	–	9.9	–	14.3	–	17.2	2.0[a]	24)

[a] Value of Φ suggested by respective authors
[b] Calculated using the theory of Flory and Orifino[56)]

Table 5. Poly(1-butene)

Tacticity	Solvent	Exp. temp. °C	θ temp. °C	$K_\theta \times 10^4$ Direct determination	(F-F)	(S-F)	(K-S)	(B)	C_∞	$\Phi \times 10^{21}$	Ref.
Atactic	n-Nonane	35	–	21^b	–	–	–	–	11.8	2.1^a	26)
	Anisole	83	83	10.81	–	–	–	–	6.7	2.5	83)
	Isoamyl acetate	23	23	11.3	–	–	–	–	6.9	2.5	83)
	Phenetole	61	61	10.5	–	–	–	–	6.6	2.5	83)
	Toluene	–46	–46	13.3	–	–	–	–	7.7	2.5	83)
	Benzene	30	–	–	–	–	11	–	6.8	2.5	83)
	Toluene	30	–	–	–	–	11	–	6.8	2.5	83)
	Decalin	30	–	–	–	–	11	–	6.8	2.5	83)
	Decalin	100	–	–	–	–	10.5	–	6.6	2.5	83)
	Tetralin	100	–	–	–	–	10.0	–	6.4	2.5	83)
Isotactic	n-Nonane	80	–	30^b	–	–	–	–	14.9	2.1^a	26)
	Cyclohexane 69 / n-Propanol 31	35	35	24.7	–	–	–	–	10.7	2.87^a	84)
	Cyclohexane	35	–	–	24.1	25.0	24.1	–	10.6	2.87^a	84)
	Cyclohexane 90 / n-Propanol 10	35	–	–	24.1	25.0	24.1	–	10.6	2.87^a	84)
	Cyclohexane 80 / n-Propanol 20	35	–	–	24.1	25.0	24.1	–	10.6	2.87^a	84)
	Cyclohexane 70 / n-Propanol 30	35	–	–	24.1	25.0	24.1	–	10.6	2.87^a	84)
	Cyclohexane 65 / n-Propanol 35	35	–	–	24.1	25.0	24.1	–	10.6	2.87^a	84)

a Value of Φ as suggested by the respective authors
b Calculated from measurements in good solvents using the method of Flory and Orifino[56]

relations have been developed. The theory behind these techniques was the subject of an extensive review by Kurata and Stockmayer[34]. In addition, Cowie[48] has published a critical review of several of the techniques. More recently, Rao et al.[49] published a review in which they tested many of the commonly used relationships.

Basically, all techniques involve taking into account the effect of long range interactions which alter the dimensions of the polymer by a factor α^{50}. As a result, the root-mean-square end-to-end distance, $\langle \bar{r}^2 \rangle^{1/2}$, in a thermodynamically good solvent may be expressed by:

$$\langle \bar{r}^2 \rangle^{1/2} = \alpha \langle \bar{r}_0^2 \rangle^{1/2} \tag{6}$$

where $\langle \bar{r}_0^2 \rangle^{1/2}$ is the root mean square end-to-end distance unperturbed by long range interactions. The intrinsic viscosity, $[\eta]$, which represents a volume, should therefore vary as α^3, and can be expressed as:

$$[\eta] = \alpha^3 [\eta]_\theta \tag{7}$$

Table 6. Polypropylene

Tacticity	Solvent	Exp. temp. °C	θ temp. °C	$K_\theta \times 10^4$ Direct determination	(F-F)	(S-F)	(K-S)	(B)	C_∞	$\Phi \times 10^{-21}$	Ref.
Atactic	1-Chloro naphthalene	74	74	18.2	–	–	–	–	8.18	2.1[a]	[17]
	Cyclohexane	92	92	17.2	–	–	–	–	7.86	2.1[a]	[17]
	Isoamyl acetate	34	34	16.8	–	–	–	–	6.12	2.1[a]	[17]
	Diphenyl ether	153	153	12.0	–	–	–	–	6.20	2.1[a]	[17]
	? Decalin	135	--	–	–	–	12.5	–	5.6	2.5	[13, 14]
	Isoamyl acetate	34	34	16.8	–	–	–	–	6.9	2.5	[91]
	Isobutyl acetate	58	58	15.8	–	–	–	–	6.7	2.5	[91]
	Diphenyl	129	129	12.8	–	–	–	–	5.8	2.5	[91]
	Diphenyl ether	146	146	12.5	–	–	–	–	5.7	2.5	[91]
Syndio-tactic	Isoamyl acetate	45	45	17.2	–	–	–	–	6.9	2.5	[16]
	Toluene	30	–	–	–	16.4	–	–	6.6	2.5	[16]
	Heptane	30	–	–	–	16.4	–	–	6.6	2.5	[16]
	Decalin	135	–	–	–	11.2	–	–	5.2	2.5	[16]
Isotactic	Diphenyl ether	145	145	13.2	–	–	–	–	7.48	2.1[a]	[17]
	Diphenyl ether	145	145	9.4	–	–	–	–	4.6	2.5	[77]
	Diphenyl ether	145	145	6.6	–	–	–	–	4.73	2.5	[85]
	Diphenyl	125.1	125.1	15.2	–	–	–	–	5.80	2.87[a]	[86]
	Diphenyl ether	142.8	142.8	13.7	–	–	–	–	5.41	2.87[a]	[86]
	Dibenzyl ether	183.2	183.2	10.6	–	–	–	–	4.56	2.87[a]	[86]
	Diphenyl	125	125	14.1	–	–	–	–	6.2	2.5	[91]
	Diphenyl ether	143	143	13.0	–	–	–	–	5.9	2.5	[91]

[a] Value of Φ suggested by the respective authors

where $[\eta]_\theta$ is the intrinsic viscosity in a theta solvent. Combining Eqs. 3 and 7 one obtains:

$$[\eta] = K_\theta M^{1/2} \alpha^3 \tag{8}$$

To obtain the unperturbed parameter, K_θ, one must choose an appropriate relation for the expansion factor, α. Several relations have been proposed, those of Fox and Flory[31], (F-F), [Eq. 9]; Stockmayer and Fixman[51], (S-F), [Eq. 10]; Kurata and Stockmayer[34], (K-S), [Eq. 11]; and Berry[52], (B), [Eq. 12], have been the most thoroughly tested, although several others have been proposed[34, 48, 53, 54, 55, 56].

$$\alpha^5 - \alpha^3 = 2\,C_M\,(1/2 - \chi)\,M^{1/2}$$

$$\text{where; } C_M = \left(\frac{27}{2^{5/2}}\,\pi^{3/2}\right)\left(\bar{v}^2/V_1 N_A\right)\left(M/\bar{r}_0^2\right)^{3/2} \tag{9}$$

$\bar{v}$ = partial specific volume of solute

N_A = Avogadro's number

V_1 = molar volume of solvent

and; χ is a parameter related to polymer-solvent interaction

$$\alpha^3 = 1 + 2z \qquad (10)$$

Table 7. Poly(1-Pentene)

Tacticity	Solvent	Exp. temp. °C	θ temp. °C	$K_\theta \times 10^4$ Direct determination	(F-F)	(S-F)	(K-S)	(B)	C_∞	Φ $\times 10^{-21}$	Ref.
Isotactic	Isoamyl acetate	31.5	31.5	12.3	–	–	–	–	9.5	2.4[a]	[87]
	2-Pentanol	62.4	62.4	12.1	–	–	–	–	9.1	2.5	[88]
	i-Butyl acetate	32.5	32.5	12.03	–	–	–	–	9.1	2.5	[38]
	Phenetole	64	64	11.3	–	–	–	–	8.7	2.5	[38]
	Anisole	85	85	10.60	–	–	–	–	8.3	2.5	[38]
	Diphenyl methane	121	121	9.77	–	–	–	–	7.9	2.5	[38]
	Phenyl ether	149	149	9.77	–	–	–	–	7.9	2.5	[38]
	Toluene	30	–	–	12.1[b]	13.6[b]	14.0[b]	11.7[b]	9.5	2.5	
Atactic	i-Butyl acetate	32.5	32.5	10.00	–	–	–	–	8.0	2.5	[38]
	Phenetole	64	64	9.77	–	–	–	–	7.9	2.5	[38]
	Anisole	85	85	9.86	–	–	–	–	7.9	2.5	[38]
	Phenyl ether	149	149	9.44	–	–	–	–	7.7	2.5	[38]

[a] Value of Φ suggested by the respective authors
[b] As shown in Figs. 2–5; calculated by the authors from data in Ref. [38]

Table 8. Poly(methyl acrylate)

Tacticity	Solvent	Exp. temp. °C	θ temp. °C	$K_\theta \times 10^4$ Direct determination	(F-F)	(S-F)	(K-S)	(B)	C_∞	Φ $\times 10^{-21}$	Ref.
Atactic	Toluene	35	–	–	6.5	6.6	6.6	–	–	2.5[a]	[92]
	Benzene	35	–	–	6.7	–	–	–	–	2.5[a]	[92]
	2-methyl cyclohexanol	56.5	56.5	6.8	–	–	–	–	–	2.5[a]	[30]
	Isoamyl acetate	61.7	61.7	6.8	–	–	–	–	–	2.5[a]	[30]
	Isoamyl acetate	61.7	61.7	7.3[b]	–	–	–	–	–	–	
	MEK/iso-propanol (1/1)	27.5	27.5	7.2	–	–	–	–	–	2.5[a]	[30]
	MEK/iso-propanol (1/1)	27.5	27.5	5.44	–	–	–	–	–	2.1[a]	[93]

[a] Value of Φ suggested by the respective authors
[b] Obtained from light scattering measurement under θ conditions

Table 9. Experimental values of the temperature coefficient for PMMA

Tacticity	d ln C ∞/dT		Ref.
	Sign ($\pm$)	Value (K^{-1})	
Atactic		0	[68]
Syndiotactic	+	1.4×10^{-3}	[70]
	+	2.4×10^{-3}	[73]
	+	4.0×10^{-3}	[90]
Isotactic	−	-2.3×10^{-3}	[70]

where; $Z = (3/2\,\pi)^{3/2}\,\beta\,(\overline{r}_0^2/M)^{3/2}M^{1/2}$ and; β is the "Binary cluster integral"[34]

$$\alpha^3 - \alpha = 4/3\,z\,g(\alpha) \tag{11}$$

where; $g(\alpha) = 8\,\alpha^3\,(3\,\alpha^2 + 1)^{-3/2}$

$$\alpha^3 = 2 + .325\,z \tag{12}$$

where; $2 < z < 11$.

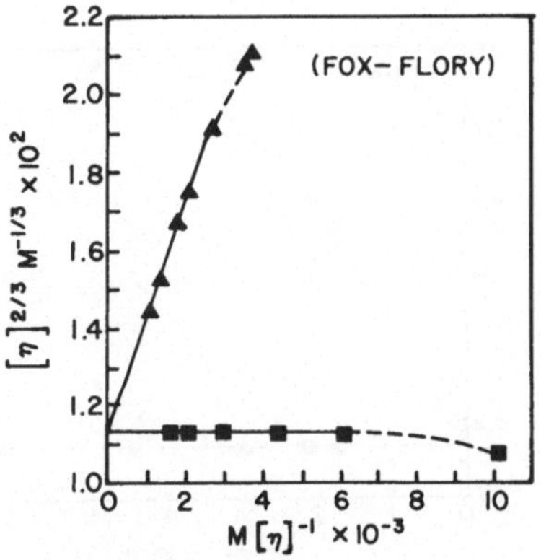

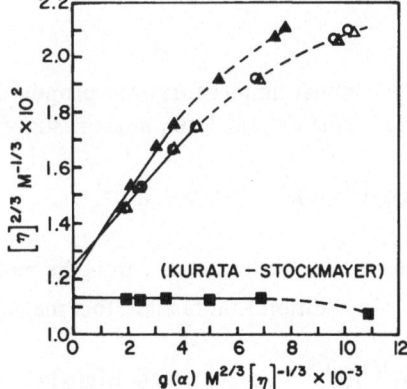

Fig. 2. Fox-Flory plot for isotactic poly(pentene-1)[38]; (▲) toluene at 30 °C; (■) i-butyl acetate at 32.5 °C, θ temperature

Fig. 3. Kurata-Stockmayer plot for isotactic poly(pentene-1)[38]. Shown in the plot are three successive approximations for $g(\alpha)$ as suggested by Kurata and Stockmayer. The values are shown to converge rapidly. (▲) toluene at 30 °C, first approximation; (△) toluene at 30 °C, second approximation; (○) toluene at 30 °C, third approximation; ■ i-butyl acetate at 32.5 °C, θ temperature

On combining Eq. 7 with each of Eqs. 9–12, relationships were obtained[31, 34, 34a, 51, 52] whereby the value of K_θ can be graphically evaluated if $[\eta]$ and M are known. Flory and Fox[31] were the first to put forward such a method; their Eq. was

$$[\eta]^{2/3}/M^{1/3} = K_\theta^{2/3} + K_\theta^{5/3} C_T (M/[\eta]) \tag{9A}$$

where; $C_T = 2\psi_1 C_M (1 - \theta/T) = (\alpha^5 - \alpha^3)/M^{1/2}$ and; ψ_1 is an entropy parameter.

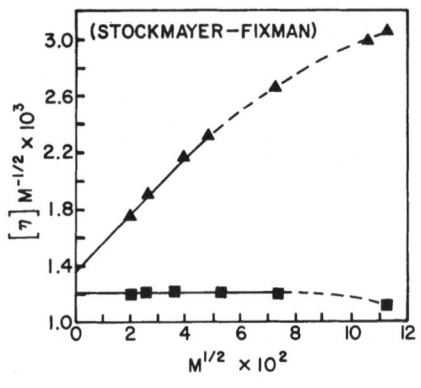

Fig. 4. Stockmayer-Fixman plot for isotactic poly(pentene-1)[38]; (▲) toluene at 30 °C; (■) i-butyl acetate at 32.5 °C, θ temperature

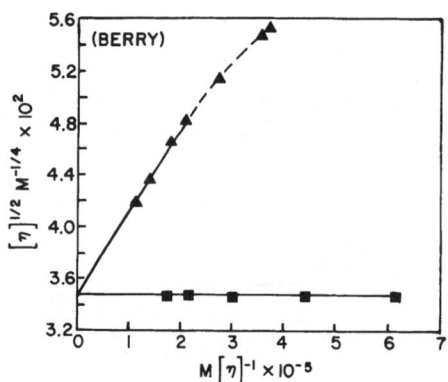

Fig. 5. Berry plot for isotactic poly(pentene-1)[38]; (▲) toluene at 30 °C; (■) i-butyl acetate at 32.5 °C, θ temperature

A much simpler form was proposed by Burchard[34a], and later by Stockmayer and Fixman[51]; the latter authors suggested the Eq.:

$$[\eta]/M^{1/2} = K_\theta + .51 \Phi B M^{1/2} \tag{10A}$$

where; $B = \beta/c^2 m^2$ and; c m is the molar weight of a chain segment.

In addition Kurata and Stockmayer[34] proposed:

$$[\eta]^{2/3}/M^{1/3} = K_\theta^{2/3} + .363 \Phi B[g(\alpha) (M^{2/3}/[\eta]^{1/3})] \tag{11A}$$

While Berry[52] proposed:

$$([\eta]/M^{1/2})^{1/2} = K_\theta^{1/2} + .42\,K_\theta^{3/2}\,B(\overline{r}_0^2/M)^{-3/2}\,(M/[\eta]) \tag{12A}$$

Typical plots for Eqs. 9A–12A are shown in Figs. 2–5. In each case values of K_θ are obtained from the intercept of the appropriate plot. It is clear that the linear relationships proposed, Eqs. 9A–12A, do not hold at high molecular weights. As a result, values of K_θ must be obtained by extrapolation of the linear portion of the curves. The data used to obtain these plots, (Figs. 2–5), is from a study by Moraglio and Gianotti[38] on isotactic and atactic PPT. This data had not been previously used to estimate unperturbed dimensions by the theories presented here. Shown in the graphs are data for isotactic PPT in toluene, at 30 °C, a thermodynamically good solvent, and isobutyl acetate, at 32.5 °C, a θ solvent. Comparison of the results obtained by these four extrapolation techniques with that obtained in a θ solvent, (iso-butyl acetate) shows reasonable agreement, (see Table 7). Values of K_θ as estimated by the Stockmayer-Fixman and the Kurata-Stockmayer methods are seen to be $\sim$10% higher than those obtained directly in a theta solvent. The values from the Fox-Flory and Berry Eqs. are even closer to the experimental value; the Fox-Flory method within 1.6%; the Berry method within 4.8%. This is exceptional agreement since the values of K_θ obtained directly in θ solvents are generally accurate to $\pm 2\%$. Deviations from the experimental values has been attributed in part to the fact that the semi-empirical parent Eqs., Eqs. 9–12, are only valid over a limited range of α values[48]. As a result, it has been suggested[48] that these estimation methods be used to compliment one another.

Values of K_θ derived from Eqs. 9A–12A from measurements at temperatures greater than theta, as well as the experimental values determined at the theta temperature, for several polymer-solvent systems, are given in Tables 1–8. It is not our purpose to critically review the various techniques for obtaining K_θ from data in non-theta solvents, as this problem has been considered elsewhere[48, 49]. However, data analysis does indicate that the agreement between the experiment and calculation is generally within 8%, and often the agreement is better. This is of prime importance to those interested in stereoregular polymers since estimations of K_θ in good solvents are far easier to obtain than the corresponding values determined in theta solvents.

2. Stereoregular Systems

Estimations of the unperturbed dimensions of several polymers[48, 57–59] have been obtained by measurements in good solvents using the extrapolation methods discussed in the previous Sect. However, relatively few have been concerned with the unperturbed dimensions of polymers of known stereoregularity. The data available on stereoregular polymers is given in Tables 1–8. With the exception of PMMA, data is scant for determinations of unperturbed dimensions in non-θ solvents. For PMMA the agreement is excellent among non-theta estimation techniques discussed here. For other polymers where estimates have been attempted, agreement is also satisfactory. For example, the value obtained by the Stockmayer-Fixman method for syndiotactic PP and that obtained by direct measurement at the θ temperature are in reasonable agreement. In addition, the data available for atactic PMA (Table 8), shows good agreement between the values of K_θ obtained by the Fox-Flory, Stockmayer-Fixman, and Kurata-Stockmayer methods, and those obtained by direct measurements at the θ temperature. Also, the limited data available for isotactic PPT as interpreted here, shows good agreement with those values determined in θ solvents at the same temperature. In general, values of K_θ obtained from the estimation methods discussed here, have thus been shown to give good agreement with the values as determined directly in theta solvents[48, 49].

IV. Data Analysis

A. Dependence of Dimensions on Stereoregularity

Immediately obvious, is the dependence of the unperturbed dimensions on the stereoregularity. For PMMA, the K_θ values for the isotactic form are ~50% larger than those of either the atactic or syndiotactic forms. Although only limited data is available for PB-1, indications are that the isotactic form is ~40% larger. For PIPA and PPT, the isotactic form is ~25% larger than either the atactic or syndiotactic forms. Although the isotactic content of PMS is not high, (~11% isotactic triads), the "isotactic polymer is ~10% larger than the syndiotactic form. No data is available for syndiotactic PS. However, the isotactic form does exhibit a K_θ ~25% larger than the atactic form. On the other hand, the data for PP, although considerably scattered, appears to indicate that K_θ for the syndiotactic and atactic forms is larger than that of the isotactic form by ~25%.

B. Error Analysis

Due to the errors inherent in the methods used to obtain K_θ, the values are believed to be accurate to ±10%. The observed differences of ≥25% are therefore considered real. However, a question that arises on comparing unperturbed dimensions, as ob-

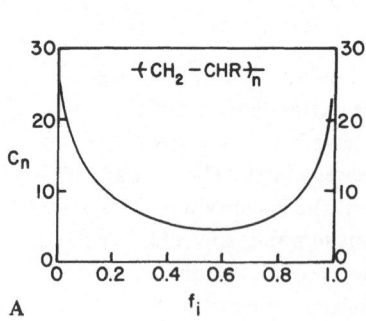

A

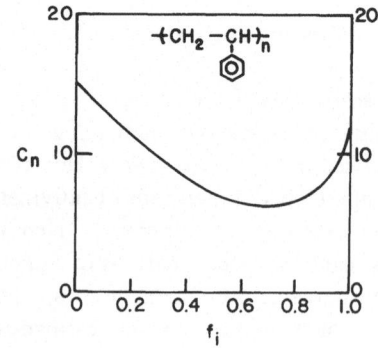

B

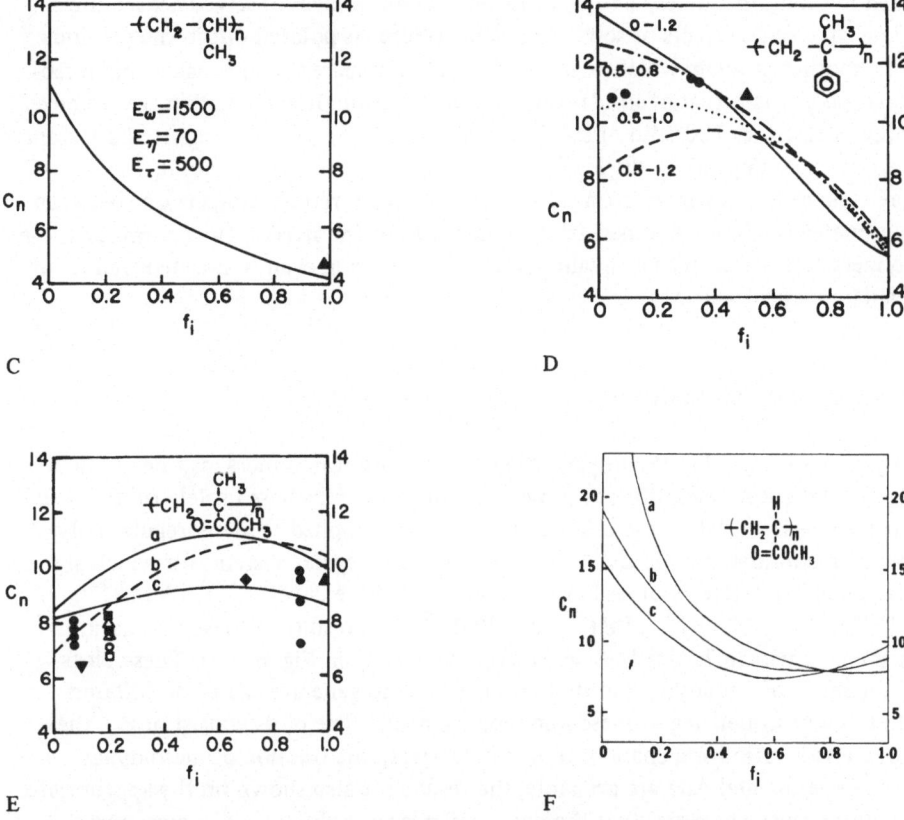

Fig. 6. A The characteristic ratio as a function of stereoregularity, calculated for a system with $R = (CH_2)_z CH_3 (z \geqslant 1)$. The parameters used are $E_T^*/RT = 2.3$, $E_\omega''/RT = 5 \pm 1$ and $\Delta\phi = 20°$, respectively. From Ref. [63]. **B** The characteristic ratios for Monte Carlo chains of 100 units each as a function of f_i, the fraction of meso dyads in the chain. The curve shown above represents results of calculations carried out with $E_T''/RT = E_\omega/RT = 5$. From Ref. [63]. **C** The characteristic ratios for Monte Carlo chains of 200 units each as a function of f_i, the fraction of meso dyads in the chain. The curve shown above represents results of calculations carried out for a temperature of 140 °C with the conformational parameters chosen as indicated, in cal. mol.$^{-1}$. The experimental values of Bovey and Heatley[85] (▲) are shown. From Ref. [65]. **D** The characteristic ratios for Monte Carlo chains of 200 units each as a function of f_i, the fraction of meso dyads in the chain. The values of E_α and E_β in that order, in kcal mol^{-1}, are marked on the curves. The experimental values of Cowie and Bywater[81] (●), and Noda et al.[80] (▲) are shown. From Ref. [67]. **E** The characteristic ratios for Monte Carlo chains of 200 units each as a function of f_i, the fraction of meso dyads in the chain. Curves are shown for (a) $E_\alpha = 1.0$, $E_\beta = -0.6$, and $\theta = 58°$; (b) $E_\alpha = 1.2$, $E_\beta = -0.6$, and $\theta' = 58°$; (c) $E_\alpha = 1.2$, $E_\beta = -0.2$, and $\theta' = 56°$, energies being in kcal mol^{-1}. The experimental results of various authors are represented by points as follows: Katime et al. (▲)[75], Katime and Roig (▲)[74], Sakurada et al. (●)[70], Fox (■)[68], Fox (■)[68], Krause and Cohn-Ginsburg (♦)[76], Chinai and Valles (△)[89], Vasudevan and Santappa (□)[69]. From Ref. [44]. **F** The characteristic ratios for Monte Carlo chains of 200 units each, plotted as a function of f_i, the fraction of meso dyads. Curve (a) $E = -0.5$; curve (b) $E = -0.3$; and curve (c) $E = -0.1$. $E_W = 2.8$; $E_{W'} = 2.2$; $E_{W''i} = 1.6$; $E_{W''d} = 1.0$; $E_p = 0.3$; $T = 300$ K. All energies are in kcal mol^{-1}.

The experimental values for atactic PMA given in Table 8 are estimated to correspond to 49–56% isotactic dyads, as estimated from the work of Matsuzaki et al. [94]

tained by solution studies, is to what extent the differences observed are manifes-
tations of specific effects of solvent or temperature, as pointed out in the previous
Sect. Another problem which arises on trying to compare the dimensions of stereoregular polymers is that of insufficient characterization. It is likely that we are never
dealing with either 100% iso or syndiotactic forms. Consequently reported differences
between forms depend on the level of stereoregularity. For most data in Tables 1—7,
there is no exact information on stereoregularity. As a result, comparisons between
experimentally observed unperturbed dimensions of the stereoregular forms of these
polymers must be deemed as qualitative. As later shown, even trends deduced from
such data may be spurious.

V. Statistical Calculations

Flory[60] has shown that the unperturbed dimensions of polymers may be predicted
through statistical evaluations of molecular conformations based solely on the rotational isomeric model. These calculations have been applied to stereoregular polymers with considerable success[61]. Among the stereoregular systems for which statistical conformational studies have been attempted are PMA[62, 62a], PMMA[44],
PS[63, 64], PP[65, 66], PPT[63], PBT[63], and PMS[67]. The results of these calculations
for the polymers in Tables 1—8 are shown graphically in Fig. 6A—F. These plots
are the result of employing Monte Carlo methods to generate chains of 200 units
with random sequencing of meso and racemic dyads. The plots consist of C_n, the
characteristic ratio for a chain of length n, versus f_i, the fraction of meso dyads.
Where experimental data are available, the results are also shown on the appropriate
plot. It is obvious however that the data available for polymers of known stereoregularity is insufficient to offer a sensitive test of these theoretical results. However, one interesting feature, as pointed out by Flory et al.[44], is the fact that for all
vinyl polymers, $\{CH_2-CHR\}$, the theoretical curve has a convex shape, whereas
for all vinylidene polymers, $\{CH_2-CR_1R_2\}$ where $R_1 \neq R_2$, the theoretical curve
has a concave shape. This indicates that for the vinylidene polymers a few percent
stereoirregularity results in an extension of the chain, whereas for the less sterically
hindered vinyl chain a few percent stereoirregularity results in a decrease in the
dimensions.

VI. Discussion

Much has been learned from both theoretical and experimental studies of the conformations of polymer chains. It is obvious however that one cannot advance without other. That is, good sound theoretical studies depend entirely on good experimental studies, and vice versa. In the case of the effect of stereoregularity on the
unperturbed dimensions, it is obvious that considerable experimental work is needed
in order to substantiate the theoretical results. However, advancements in both synthesis and characterization should result in a considerable improvement in our understanding in this area.

VII. References

1. Natta, G.: J. Polym. Sci. *16*, 143 (1955)
2. Danusso, F., Moraglio, G.: J. Polym. Sci. *24*, 161 (1957)
3. Natta, G., Danusso, F., Moraglio, G.: Makromol. Chem. *20*, 37 (1956)
4. Peaker, F. W.: J. Polym. Sci. *22*, 25 (1956)
5. Trossarelli, L., Campi, E., Saini, G.: J. Polym. Sci. *35*, 205 (1959)
6. Ang, F.: J. Polym. Sci. *25*, 126 (1957)
7. Ang, F., Mark, H.: Monatsh. Chem. *88*, 427 (1957)
8. Krigbaum, W. R., Carpenter, D. K., Newman, S.: J. Phys. Chem. *62*, 1586 (1958)
9. Krigbaum, W. R., Carpenter, D. K.: J. Phys. Chem. *59*, 1166 (1955)
10. Outer, P., Carr, C. I., Zimm, B. H.: J. Chem. Phys. *18*, 830 (1950)
11. Chinai, S. N., Scherer, P. C., Boundurant, C. W., Levi, D. W.: J. Polym. Sci. *22*, 527 (1956)
12. Notley, N. T., Debye, P. J. W.: J. Polym. Sci. *17*, 99 (1955)
13. Danusso, F., Moraglio, G.: Makromol. Chem. *28*, 250 (1958)
14. Kinsinger, J. B., Hughes, R. E.: J. Phys. Chem. *63*, 2002 (1959)
15. Parrini, P., Sebastiano, F., Messina, G.: Makromol. Chem. *38*, 27 (1960)
16. Inagaki, H., Miyamota, T., Ohta, S.: J. Phys. Chem. *70*, 3420 (1966)
17. Kinsinger, J. B., Hughes, R. E.: J. Phys. Chem. *67*, 1922 (1963)
18. Chiang, R.: J. Polym. Sci. *28*, 235 (1958)
19. Krigbaum, W. R., Woods, J. D.: J. Polym. Sci. A-2, *2*, 3075 (1964)
20. Krause, S., Cohn-Ginsberg, E.: Polymer *3*, 565 (1962)
21. Chiani, S. N., Guzzi, R. A.: J. Polym. Sci. *21*, 417 (1956)
22. Chiani, S. N., Matlock, J. D., Resnick, A. L., Samuels, R. J.: J. Polym. Sci. *17*, 391 (1955)
23. Tsvetkov, V. M., Skazka, V. S., Krinorucho, N. M.: Vysokomol. Soedin. *2*, 1045 (1960)
24. Wessling, R. A., Mark, J. E., Hughes, R. E.: J. Phys. Chem. *70*, 1903 (1966)
25. Mark, J. E., Wessling, R. A., Hughes, R. E.: J. Phys. Chem. *70*, 1895 (1966)
26. Krigbaum, W. R., Kurz, J. E., Smith, P.: J. Phys. Chem. *65*, 1984 (1961)
27. Flory, P. J.: Principles of Polymer Chemistry, Ithaka, London: Cornell Univ. Press 1975
28. Zimm, B.: J. Chem. Phys. *16*, 1093 (1948)
29. Chinai, S. N., Bondurant, C. W., Jr.: J. Polym. Sci. *22*, 555 (1956)
30. Kato, T., Miyaso, K., Noda, I., Fujimoto, T., Nagasawa, M.: Macromolecules *3*, 777 (1970)
30a.Matsuda, H., Yamano, K., Inagaki, H.: J. Polym. Sci. A-2, *7*, 609 (1969)
31. Flory, P. J., Fox, T. G.: J. Amer. Chem. Soc. *73*, 1904 (1951)
32. Cowie, J. M. G.: J. Polym. Sci. C *23*, 267 (1968)
33. Ptitsyn., O. B., Eisner, Y. Y.: J. Phys. Chem. USSR. *32*, 2464 (1958)
34. Kurata, M., Stockmayer, W. H.: Fortschr. Hochpolym. Forsch. *3*, 196 (1963)
34a. Burchard, W.: Makromol. Chem. *50*, 20 (1961)
35. Tsuruta, T, O'Driscoll, K. F.: Strukture and Mechanism in Vinyl Polymerization, New York: Dekker 1969
36. Orofino, T. A., Mickey, J. W., Jr.: J. Chem. Phys. *38*, 2512 (1963)
37. Dondos, A., Benoit, H.: Macromol. *4*, 279 (1971)
38. Moraglio, G., Gianotti, G.: Europ. Polym. J. *5*, 781 (1969)
38a.Bazuaye, A., Huglin, M. B.: Polymer *20*, 44 (1979)
39. Lifson, S., Oppenheim, I.: J. Chem. Phys. *33*, 109 (1960)
40. Bates, T. W., Ivin, K. J.: Polymer *8*, 263 (1967)
41. Ivin, K. J., Ende, H. A., Meyerhoff, G.: Polymer *3*, 129 (1962)
42. Flory, P. J., Crescenzi, V., Mark, J. E.: J. Amer. Chem. Soc. *86*, 146 (1964)
43. Bianchi, U.: J. Polym. Sci. A *2*, 3083 (1964)
44. Sundararajan, P. R., Flory, P. J.: J. Amer. Chem. Soc. *96*, 5025 (1974)
45. Lath, D., Bohdanecky, M.: J. Polym. Sci. P. L. *15*, 559 (1977)
46. Flory, P. J., Hoeve, C. A. J., Ciferri, A.: J. Polym. Sci. *34*, 337 (1959)
47. Ciferri, A.: J. Polym. Sci. A, *2*, 3089 (1964)
48. Cowie, J. M. G.: Polymer *7*, 487 (1966)

49. Roa, T. V. R., Swamy, K. N.: Z. Physik Chem. *257*, 17 (1976)
50. Flory, P. J.: J. Chem. Phys. *17*, 303 (1949)
51. Stockmayer, W. H., Fixman, M. J.: J. Polym. Sci. C *1*, 137 (1963)
52. Berry, G. C.: J. Chem. Phys. *46*, 1338 (1967)
53. Yamakawa, H., Tanaka, G.: J. Chem. Phys. *47*, 3991 (1967)
54. Alexandrowicz, Z.: J. Polym. Sci. C *23*, 301 (1968)
55. Norisuye, T., Kawahara, K., Teramoto, A., Fujita, H.: J. Chem. Phys. *49*, 4330 (1968)
56. Orofino, T. A., Flory, P. J.: J. Chem. Phys. *26*, 1067 (1957)
57. Lin, F. C., Stivala, S. S., Biesenberger, J. A.: J. Appl. Polym. Sci. *17*, 3465 (1973)
58. Katime Amashta, I. A., Sanchez, G.: Europ. Polym. J. *11*, 223 (1975)
59. Hadjichristidis, N., Devaleriola, M., Desreux, V.: ibid. *8*, 1193 (1972)
60. Flory, P. J.: Statistical Mechanics of Chain Molecules, New York: Interscience 1969
61. Flory, P. R., Sundararajan, P. R., De Bolt, L. C.: J. Amer. Chem. Soc. *96*, 5015 (1974)
62. Yoon, D. Y., Suter, V. W., Sundararajan, P. R., Flory, P. J.: Macromolecules *8*, 784 (1975)
62a. Ojalvo, E. A., Saiz, E., Masegosa, R. M., Hernandez-Fuentes, I.: Macromolecules *12*, 865 (1979)
63. Akihiro Abe: Polym. J. *1*, 232 (1970)
64. Yoon, D. Y., Sundararajan, P. R., Flory, P. J.: Macromolecules *8*, 776 (1975)
65. Suter, U. W., Flory, P. J.: Macromolecules *8*, 765 (1975)
66. Biskup, U., Cantow, H. J.: Macromolecules *5*, 546 (1972)
67. Sundararajan, P. R.: Macromolecules *10*, 623 (1977)
68. Fox, T. G.: Polymer *3*, 111 (1962)
69. Vasudevan, P., Santappa, M.: J. Polym. Sci. A-2, *9*, 483 (1971)
70. Sakurado, I., Nakajima, A., Yoshizaki, D., Nakamae, K.: Kolloid-Z. Z. Polym. *186*, 41 (1962)
71. Gundiah, S., Mohite, R. B., Kapur, S. L.: Makromol. Chem. *123*, 151 (1969)
72. Hamori, E., Pusinowski, L. R., Sparks, P. G., Hughes, R. E.: J. Phys. Chem. *69*, 1101 (1965)
73. Schulz, G. V., Wunderlich, W., Kriste, R.: Makromol. Chem. *75*, 22 (1964)
74. Katime, I. A., Roig, A.: Ann. Quimica *69*, 1217 (1973)
75. Katime, I. A., Roig, A., Leon, L. M., Montero, S.: Europ. Polym. J. *13*, 59 (1977)
76. Krause, S., Cohn-Ginsburg, E.: J. Phys. Chem. *67*, 1479 (1963)
77. Brandrup, J., Immergut, E. H.: Polymer Handbook, New York: Interscience 1966
78. Krigbaum, W. R., Flory, P. J.: J. Polym. Sci. A *2*, 4533 (1951)
79. Fox, Jr., T. G., Flory, P. J.: J. Phys. Chem. *73*, 1915 (1951)
80. Noda, I., Mizutani, K., Kato, T., Fujimoto, T., Nagasawa, M.: Macromolecules *3*, 787 (1970)
81. Cowie, J. M. G., Bywater, S.: J. Polym. Sci. A-2, *6*, 499 (1968)
82. Okamura, S., Higashimura, T., Imanishi, Y.: Chem. High Polym. *16*, 244 (1959)
83. Moraglio, G., Gianotti, G., Danusso, F.: Europ. Polym. J. *3*, 251 (1967)
84. Satyanarayana Sastry, K., Patel, R. D.: ibid. *5*, 79 (1969)
85. Heatley, F., Salovey, R., Bovey, F. A.: Macromolecules *2*, 619 (1969)
86. Nakajima, A., Saijo, A.: J. Polym. Sci. A-2, *6*, 735 (1968)
87. Moraglio, G., Brzinki, J.: J. Polym. Sci. B, *2*, 1105 (1964)
88. Mark, J. E., Flory, P. J.: J. Amer. Chem. Soc. *87*, 1423 (1965)
89. Chinai, S. N., Valles, R. J.: J. Polym. Sci. *39*, 363 (1959)
90. Moore, W. R., Fort, R. F.: J. Polym. Sci. Part A *1*, 929 (1963)
91. Moraglio, G., Gianotti, G., Bonicelli, U.: Europ. Polym. J. *9*, 623 (1973)
92. Karunakarn, K., Santappa, M.: J. Polym. Sci. A-2, *6*, 713 (1968)
93. Takahashi, A., Kawai, T., Kagawa, I.: Nippon Kagaku Zasshi *83*, 14 (1962)
94. Matsuzaki, K., Uryu, T., Ishida, A., Takeuchi, M., Ohki, T.: J. Polym. Sci. A-1, *5*, 2167 (1967)

Received February 22, 1980
H.-J. Cantow (editor)

Reinforcement of Rubber by Carbon Black

Zvi Rigbi

Department of Chemical Engineering, Technion Israel Institute of Technology,
Haifa, Israel

The behaviour of carbon black upon addition to elastomers to increase the strength, particularly the abrasion resistance and the tear strength, of the crosslinked product is reviewed. It is found that the overwhelmingly greater portion of all studies on carbon black/elastomer interactions deals with properties at extensions much lower than those obtaining at rupture. These are developed to show that the interactions are time- and temperature-dependent.

A theory is developed which makes it possible to explain the phenomena of stress softening, the increase of strength and other properties as a result of the use of carbon black, as well as the effect of temperature and strain-rate upon them.

Table of Contents

1. Introduction

The mechanism by which carbon black reinforces elastomers is one of the most interesting problems of modern technology, and still a subject of much speculation. The effect itself was discovered by S. C. Mote of the India-Rubber, Gutta-Percha and Telegraph Works Co. of Silvertown, near London, as a result of an effort to improve the properties of rubber even more than was possible by zinc oxide. The (American) Goodrich Company purchased certain manufacturing rights from the Silvertown company in 1912, and by 1915, the inclusion of carbon black in rubber compounds of high quality had become general. The most important effect observed upon introduction of carbon black into rubber was the vast improvement in abrasion resistance. As long as natural rubber was the unique elastomer in general use, other effects of carbon black, although by no means insignificant, were not of the same great importance, with the possible exception of tear resistance.

With the outbreak of war in 1939 and the consequent shortage of natural rubber in the industrial West, the introduction of the copolymer of butadiene and styrene, then known as Buna or GR-S, as an almost total replacement for natural rubber, would not have been possible without carbon black. It may be stated, with little likelihood of contradiction, that no other product exists which contributes as much strength and abrasion resistance to noncrystallizing rubbers, while maintaining to a large extent their desirable elastic properties, as does carbon black.

In spite of the tremendous consumption of carbon black, mainly in tires, and the steady progress made in improving its quality, the reasons for its unique behaviour are still largely a matter for speculation and debate. The subject is enormously complex, as may be gathered from a survey of the literature. Many studies have been published describing one aspect or another of the behaviour of carbon black (or carbon blacks, since there are many grades available) and these are still appearing. Some excellent papers and books survey and analyse the available information, such as those by Studebaker[90], Kraus[49−51], Donnet and Voet[22] and Medalia[63].

There is a certain polytomy in defining the term reinforcement: until the advent of synthetic rubber, the only practical interest in carbon black was concerned with abrasion resistance and tear strength, the two being possibly related. In the early literature on the subject, most of the studies reported upon were carried out in simple tension, because it was not known at the time how to perform valid tests for abrasion and tear resistance. (It may be noted that to this day, no research worker is really happy with tests for abrasion resistance). However, it is now generally agreed that reinforcement is an effect by which abrasion resistance, tear strength and tensile strength are simultaneously improved, the first to a very considerable degree. These properties are usually accompanied by increased hardness, tensile product (energy density to break) or improvements in other properties, but such effects are not considered to be necessarily indicative of a capacity for reinforcement by the filler involved. On the other hand, reinforcement is almost always observed to occur together with some undesirable effects in the rubber compound, such as increased relaxation and creep rates, compression or tensile set and hysteresis. The achievement of a proper balance between these undesirable contributions of a reinforcing filler and even a minor improvement of abrasion resistance, for example, is one of the arts of the rubber compounder and the reason for the availability of a large number of grades of carbon black. One outcome of the hypothesis proposed by the author of this review[75-77] is the dual nature of the effect of carbon black, whereby the improvements desired in the behaviour of a rubber compound cannot be obtained without parallel sacrifices. Sakurai et al.[91] have stressed this point previously.

This treatise is not intended to be an encyclopedia of carbon black technology. It is a review written to analyse opinions concerning the behaviour of carbon black filled elastomers insofar as the carbon black affects the behaviour at or near failure, whether by abrasion, tear or tensile rupture. Behaviour at low or moderate extensions, which has been adequately reviewed, will not be considered unless it can be shown to bear upon reinforcement in the sense of the word implied in any dictionary definition[1]. It will be seen that a number of phenomena involved require the formulation of a specific model which shall be treated in considerable detail in the second half of this review.

The writer assumes that all the properties and parameters studied in rubber science are well understood by his readers. Test methods will not be described unless necessary.

2. Some General Comments on Carbon Black/Rubber Interactions

A number of the effects observed upon mechanically mixing carbon black with elastomers are superficially sufficiently similar to those resulting from their crosslinking that many writers have made attempts to consider the two as similar[25]. This view is now largely discounted. It may be calculated that even in a lightly cured compound, the distance between crosslinks is of the order of 3 nm. According to manufacturers data, the mean particle size of an HAF black is of the order of 30 nm, and the average distance between carbon black aggregates in a tire-tread mix is about 100 to 300 nm. These dimensions are given a graphical representation in Fig. 1. It may be concluded that even if the chemistry of the elastomer-to-black interaction were similar to that of crosslinks between chains, the overall effects observed when comparing reinforced but not otherwise crosslinked rubber with a cured gum elastomer would have to be very different indeed. Dijauw and Gent[21] have shown how carbon black affects the viscoelastic properties of SBR, and it may be seen from their work that the mixes behave very differently from cured elastomers.

In this respect, it is important to note the conclusions reached by Westlinning[99] who compared the abrasion resistance at the same modulus of two series of compounds, one in which the modulus was obtained by the use of sulfur and curatives, the other by the addition of carbon black to a single basic compound. He found that while the abrasion resistance of the latter increased up to a certain maximum, the resistance of the former fell continuously and rapidly. Also, abrasion resistance is

---

1   Reinforce, *v. t.,* – "Strengthen or support by additional men or material, or by increase of number, quantity, size, thickness etc." (Oxford). "To strengthen with new force, aid, material or support". (Merriam-Webster). "To strengthen by the addition of something". (English Larousse)

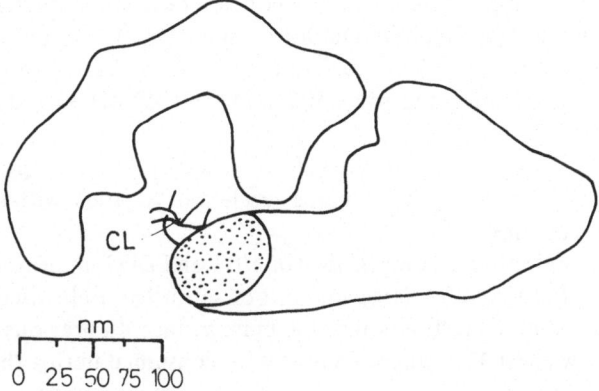

Fig. 1. Sketch showing two aggregates of N-375 carbon black about 300 nm apart. Lower aggregate with a primary particle approximately 30 nm across, and two crosslinks marked CL about 3 nm apart on a rubber molecule. Outlines freely enlarged from middle portion of right-hand edge of Fig. 5 in Ref.[63]

known to fall rapidly with increased modulus as the result of the use of nonreinforcing fillers.

Of importance to later conclusions are observations made by a number of workers. Hobden and Jellinek[43] studied the adsorption of polymer from solution by porous carbons and found that it is a process which requires considerable time. This was also observed by Kraus and Dugone[52] and Kraus and Gruver[54] who found that as the surface area of the blacks increased, longer periods were required to achieve equilibrium, and that absorption rates decrease with increasing molecular weight. Gessler[29] and Kolthoff, Gutmacher and Kahn[48] studied adsorption of polymer from solutions of polyisobutylene and SBR, and observed that low molecular fractions were first absorbed, and that this was gradually displaced by fractions of higher molecular weight.

In discussing adsorption from solutions, it is necessary to consider the presence of the solvent itself. Rivin[71] showed that toluene is adsorbed onto the carbon black surface, and is progressively removed as temperature is increased, some toluene being retained till about 165 °C. When studying adsorption from solution, it therefore appears possible that the solvent is adsorbed onto active sites, this being displaced by polymer fractions of gradually increasing molecular weight, the rates at which these exchanges probably increasing with temperature.

Unpublished work by Lambert[57] dealt with adsorption of SBR from mixed high (SBR 1500) and low molecular weight polymers ("Flosbrene") in the absence of solvent. As in the case of solutions, it was found that the low molecular weight fraction was first adsorbed, but that over a period of weeks, this was progressively displaced by high molecular weight material.

3. Immobilization of Elastomer Macromolecules by Brass

Early in his investigations, the writer attempted to eliminate the possibility of relative movement between elastomer chains and filler particles by using brass powder as a model filler. Brass, as is well known, may be vulcanized to rubber to give very strong temperature-insensitive bonds. The mechanism involved was studied by Gurney[33] and Buchan and Rae[14] amongst others, and was recently investigated in great detail by Van Ooij[96]. Rigbi[79] used brass powder approximately 8000 nm across (compared to some 500 nm for an MT black) and obtained the following results:

a) Using dicumyl peroxide as a curative, the brass powder acted as an inert filler.

b) Using MBT as a unique accelerator together with sulfur, a very hard compound was obtained.

c) With large proportions of MBT and DPG, intermediate hardness was developed.

In (b) and (c), large amounts of sulfur had to be used, most of which was consumed in the sulfidation of the brass surface. In previous experiments, it had been shown that 17.7 parts of sulfur were consumed during the vulcanization process at the surface of 100 parts of the particular brass powder used. In the compounds studied, only the excess sulfur (3.3 phr) was involved in the crosslinking reaction with MBT. The mixed accelerator system is very much more active, and it was concluded that sulfur was much more rapidly taken up by the rubber, leaving insufficient for the sulfidation, with the result that there was less bonding. In the peroxide vulcanizate, there was of course no bonding whatsoever, and the compound was very flexible and soft, with high elongation.

When compared with compounds loaded with MT black, the compounds cured by sulfur were very much harder, very much weaker and had low values of elongation at break. These effects are generally only very slightly influenced by particle size.

Brass is certainly not an ideal material for this study. It would have been for better to use a polyisoprene grafted black such as prepared by Le Bras and Papirer[58], provided no other bonding sites were left available on the surface, and that it could be made certain that the bonded polymer could crosslink into the added rubber.

Unsatisfactory as the experimental evidence is, it is nevertheless the writers opinion that these results demonstrate that, whatever the mechanism of reinforcement may be, it cannot involve rigid, immovable bonds to the filler material.

4. Bound Rubber

The quantity of rubber insoubilized by carbon black due to adsorption from the bulk is known as bound rubber. It has long been assumed that the mechanism by which bound rubber is formed is also responsible for reinforcement in the vulcanized elastomer. This section deals with several aspects of the formation of bound rubber and its relationship to reinforcement.

Bound rubber is usually measured by the weight of elastomer remaining attached to carbon black after lengthy extraction with a low boiling solvent, such as benzene or hexane, occasionally at boiling temperature, often at room temperature. It has apparently always been assumed that given sufficient time, all soluble material would ultimately be extracted, leaving a layer which insoluble under any condition. However, the work of Rivin, Aron and Medalia[72], Rivin[71] and Sircar and Voet[85] have shown that extraction increases with temperature, and *by extrapolation,* a temperature exists which all of a simple polymer model compound[71, 72] or elastomer[85] will have been extracted. This temperature (T_m) is about 370 °C for SBR, 245° for butyl rubber and polyisoprene adsorbed on HAF black. Sircar and Voet found that the molecular weight of the adsorbed polymer had little influence upon the value of T_m thus determined. From the elastomer immobilized after milling at low temperature in order to encourage chain scission, these authors indicate that the multicontact adsorption theory[84] is not tenable except possibly at high loadings.

Sircar and Voet correctly conclude that extraction involves a spectrum of bonding energies, just as Beebe and coworkers[3], for example, determined spectra of adsorption energies for various gases. Presumably, then, extraction at T_m involves the highest adsorption energy in the interaction, which is assumed to be chemisorption. Immobilized elastomer seems to increase if the elastomer is masticated with the carbon black rather than adsorbed on to it from solution. Because free radicals are formed during mastication there may be more than one mechanism involved in adsorption. It would have been useful to know whether the T_m value for the masticated mix is also different from the value for the elastomer adsorbed from solution. This would have made it possible to estimate whether the bonding energy for the elastomer is different to that which bonds the free radicals. Papirer, Nguyen and Donnet[67] have shown that grafted polymer is also characterized by a T_m value, but that this is higher by some 80° than for polymer "bound" by mill mixing.

From the figures given by Rivin[76], it would appear that the maximum bonding energy of SBR on HAF black should be almost identical to that of p-xylene $(T_m \simeq 375 °C)$ at 0% coverage, and that of butyl rubber almost equal to the bonding energy for acetone $(T_m \simeq 250 °C)$. The latter is given by Rivin as 17 kcal/mol, but there seems to be no corresponding information in the literature for xylene.

The concept according to which elastomer is bonded by a spectrum of energies receives support from the excellent photographs by Ban and Hess[1] and by Ban, Hess and Papazian[2], in which extraction by benzene appears to have removed rubber from parts of the surface of the black, leaving a non-uniform layer on other parts. This would appear to be confirmed by Rivin's analysis of the acetone desorption curves from ISAF black, in which several activation energies were determined.

Pliskin and Tokita[70] developed a technique which makes it possible to separate what they term "truly adsorbed rubber" from other insoluble gel in rubber/black mixes. This presumably includes very highly convoluted material and lightly cross-linked elastomer resulting from a variety of causes. They then analyse their results to estimate the thickness of the layers of various elastomers adsorbed onto carbon black.

O'Brien et al.[65], Gessler[29], Smit[86], Kaufman, Slichter and Davis[46] and Schoon[82] studied the various phenomena which should result from the assumption of the presence of immobilized layers of rubber, and concluded that such do in fact exist. Schoon claims to have observed them as uncrystallizable layers of rubber, about 500 Å thick, whereas Smit calculated the thickness of the layers to be of the order of 20 Å, basing himself on dynamic measurements. This was in the range found by Ban, Hess and Papazian[2]. Kaufman et al.[46] concluded that there are two shells of elastomer surrounding each carbon black particle, the outer one being of limited mobility, the inner one completely immobile. Gessler[29] confirmed the existence of a bound rubber shell by studying the swelling of bound rubber gel. O'Brien et al.[65] confirm other observations but conclude that there is a gradual increase of mobility in the rubber with distance from the black surface.

It may be concluded from the above that a layer of polymer is adsorbed onto the surface of carbon black; that the process is one involving measurable rates; that the the polymer may be desorbed provided sufficient energy is available; and that this layer is fixed only in the direction normal to the carbon black surface, movement parallel to the surface being possible. On a different level of molecular weight, Wake[97] showed that similar behaviour was involved when C_4 and C_5 hydrocarbons were adsorbed onto carbon black.

5. Hysteresis and Reinforcement

When moderately filled rubber compounds are compared at equal volume loading, their hysteresis, heat build-up and compression set are known to increase as the reinforcing ability of the filler becomes more pronounced, provided that the state of cure is maintained at the same level. This is well demonstrated in Figs. 3 and 4 of Smit's work[86]. At higher volume loadings, this "rule" is less generally applicable, and certain fillers, such as silica, behave in an anomalous fashion due to poor dispersion. For a considerable portion of the loading range, the following rule also applies: as loading increases, so does reinforcement, and so do hysteresis losses and compression set[32]. It seems to be the opinion of some workers that there is a direct cause-and-effect relationship between an energy-loss mechanism in the reinforced rubber and the prevention or delay of crack propagation. For some time, Payne and his coworkers had been studying the problem and obtained important and interesting results. The point was made in different ways in the various publications, such as Harwood's[40] statement that " . . . in rubbers that show hysteresis strongly, only a small proportion of the applied energy is available to propagate flaws", although no hint is given on how this occurs.

In a test involving low-amplitude oscillations, hysteresis is usually measured by the out-of-phase modulus or loss as it is alternatively called, although possibly the tangent of the loss angle (tan δ) may be a better indicator. Be that as it may, Payne and Whittaker[68] observe that as carbon black loading is increased, so are the loss moduli and tan δ. Ulmer, Hess and Chirico[94] also demonstrate this dependence with a series of increasingly reinforcing blacks at constant loading. However, Chasset[16] presents evidence which purports to show that the mechanisms of relaxation operating in gum rubber, and in black reinforced rubber are identical. His conclusions are drawn from the analysis of test results over periods greater than 10 sec after the application of the load. As will be shown later, the major effects of black/elastomer interaction appear very rapidly, well under 10 sec.

It is a pity that because whiting, clay and other such fillers are known not to reinforce natural rubber, they have not been studied in the same way and to the same depth as carbon black. Nevertheless, it is within the experience of most rubber compounders that as the reinforcing effect increases (it may be better said that as

the weakening effect if the "white" filler decreases) so do the hysteresis and compression set at the same volume loading increase. Such a progression will be found
in the series: coarse ground whiting, precipitated calcium carbonate, soft clay, hard
clay, precipitated calcium silicate. Certain precipitated silicas and pyrogenic silica
certainly increase the strength and abrasion resistance of rubber, but when used in
more than very small proportions, they impart distinctly unrubberlike properties
to the rubber compound.

A very serious question arises when comparing the hysteresis losses of a series
of compounds. Is it fair to compare the compounds at equal loadings, or should the
hardness (or modulus) be kept constant by adjusting loadings? Should the maximum
stress reached during a cycle be kept constant and equal for all the compounds, or
should it be the elongation? Or should *energy* losses be compared at the same *energy* input, as was reported by Peremsky[69] and by Brennan, Dannenberg and Rigbi[11].

Mullins and Tobin[64] introduced the use of a "strain-amplification factor"[2] to
reduce the overall strain measured in a filled compound to the average strain of the
rubber phase in the compound. There are very serious doubts whether this factor
can apply to finite strains at all, and in fact, studies[94] have shown that at strains
as low as 10%, the Guth-Gold factor is not even approximately applicable to any
except N-990 black. At greater strains, of the order of 500% as studied by Harwood,
Mullins and Payne[39] and Harwood and Payne[40], very much greater differences
between experimental and calculated values of the Guth-Gold factor must exist.
Kraus[31] has stated much the same in different terms.

Further discussion of the questions raised above will be found in the section on
stress-softening.

Payne and Whittaker[68] list the major sources of hysteresis in filled rubbers. Two
of these, crystallization and changes in network structure, are of no interest in the
present context, as the phenomena discussed are also observed in black-loaded SBR
(a non-crystallizing rubber) and with dicumyl peroxide curing (a very stable system
of crosslinking). Breakdown of filler aggregates, while obviously a function of stress,
cannot conceivably be a function of time (which will be considered in great detail
further on). Of the remaining two sources of hysteresis listed, evidence of viscoelasticity (presumably of the rubber phase) in a material is *always* sought in hysteresis,
measured in one way or another. The real question here is whether the presence of
the filler increases the viscoelasticity, and therefore the hysteresis of the rubber phase
over that of the original volume of rubber. Whether or not the Guth-Gold factor can
be applied as it stands, it is obvious that this is only a volume effect, and therefore
cannot be the source of differences between materials. Because occluded rubber is
highly restrained in its mobility, it is not likely that it plays an important part rule

---

2 It is interesting to note that the mathematical expression used for the effect described, although
   universally credited to Eugene Guth and Otto Gold, is not identical with the one they state in
   Ref.[34]. This reference is to an abstract in which the relationship is given (using the notation
   adopted in the rubber literature) as
   $\eta_{solv} = \eta_{mix} (1 - 2.5\,\phi - 14.1\,\phi^2)$ which can be rewritten as
   $\eta_{mix} = \eta_{solv} (1 + 2.5\,\phi + 20.35\,\phi^2)$ and *not* as usually given
   $\eta_{mix} = \eta_{solv} (1 + 2.5\,\phi + 14.1\,\phi^2)$. This writer has not been able to locate the paper in which
   the first change to the currently used form appeared

in the development of hysteresis. However, Ulmer et al.[94] showed that the Guth-Gold factor can be made to fit experimental results *at low strains* by taking occluded rubber into account in the way developed by Medalia[61]. This too is almost uniquely a volume effect.

6. Stress Softening and Reinforcement

The phenomenon of stress-softening is very closely allied to that of hysteresis. It has been studied from various points of view by Kraus, Childers and Rollman[53], by Brennan, Dannenberg and Rigbi[11], and by Brennan, Jermyn and Perdigao[12], Dannenberg and Brennan[20], by Payne and coworkers[38, 39, 40, 68], by Peremsky[69] and by Rigbi[75, 76]. Payne and Whittaker[68] showed that stress softening occured not only in reinforced rubber, but also in gum rubber and materials such as full-chrome leather, poromeric leather substitutes and even a model system of knots tied in string. There is a very important difference between the mechanisms operating in leather and rubber, and the similarities in behaviour should no be allowed to obscure the differences. As will be seen, stress softening in reinforced vulcanizates is to a very large part recoverable under proper treatment, and in leather it certainly is not. Whether it develops, and if it then can be made to recover in stable, cross-linked SBR depends largely on test conditions and remains to be demonstrated.

As traditionally studied, stress softening is observed as a shift of the stress strain curve. In order to be able to study the influence of a wide number of parameters, it is necessary to find a basis for comparison which will not in itself affect stress softening in different ways when these parameters are changed. Dannenberg and Brennan[20] and Peremsky[69] chose to look upon energy in the first stress cycle as the independent variable for the comparison of stress-softening between samples of different rubbers and with varying loadings. The problem of the selection of the proper strain to apply is one of critical importance, and it is proposed to examine this now at some detail.

If it is assumed that stress-softening occurs uniquely in the rubber phase of a loaded elastomer, then a method of loading or straining should be chosen whereby the rubber in all the mixes to be compared, including any gum rubber, should undergo the same stress and strain. This is obviously impossible, since compounds vary widely in crosslink density and modulus, so that a choice must be made between stress or strain. There is no clear preference for one over the other, and it must be admitted that for all practical compounds of primary interest, the *modulus of the rubber phase* in the filled rubber is likely to be sensibly constant. The situation is of course very different if the moduli of the compounds themselves are compared. Assuming for the moment the constancy of modulus in the rubber phase for all compounds, it is necessary to enquire why the moduli of compounds at equal loading with various carbon blacks are so different. At a volume loading of 40 phr, a ratio of moduli of about 4 would be found when comparing an MT black with an ISAF black. This is due to the smaller particle size and higher structure of the ISAF black, considerations which are not included in the Guth-Gold "amplification factor" as used by Payne and coworkers and referred to in the previous section.

Payne and coworkers[39] could not have considered Medalia's correction[61, 62] for occluded rubbers. Taking this into account, calculation shows that at 80 phr loading of the various blacks, the strain amplification factor should be about 6.2 for MT black (N-990) and about 18.2 for ISAF black (N-220). On the same basis, Payne et al. would have used a factor of 4.97 for all the blacks. The analysis of their results as given in Ref.[39] and in later papers therefore loses much if not all of its significance, and their conclusion based on this analysis becomes untenable. In particular, the claim that stress softening occurs principally in the rubber phase of a compound cannot be sustained, unless further justified. The strain amplification factor gives an average value for the influence of the filler on the strain; it does not give a picture of the very high strains which would develop locally near the surface of the black. As we know stress-softening increases rapidly with strain even in gum rubbers, Payne's method of calculation unwittingly increases the measured overall stress softening to the extent that it appears to develop principally in the rubber phase of the compound.

In the investigation of stress softening, Kraus, Childers and Rollman[53] varied temperatures and rates of loading and showed that in general, those conditions which tended to increase the stiffness of the vulcanizate increased the stress softening. They also tried to relate these parameters by a WLF shift-factor. Rigbi[75] studied the behaviour of compounds at a wider range of temperatures and very much wider range of strain rates (up to 6600% per sec). He was able to show while Kraus's general principle applied at moderate temperatures for low strain-rates, softening tended to *decrease* with strain rate below a certain minimum at low temperature. Also, as the temperature was decreased, so did the influence of prestressing at given rates decrease. This is shown in Fig. 2, 3 and 4, redrawn from Ref.[75].

The compounds used were as follows:

	1403 – 1	1403 – 5
High cis polybutadiene	100 phr	– phr
SBR 1500	–	100
Zinc oxide	3	3
Stearic acid	1.5	1.5
Flexamine antioxidant	1	1
Sulfur	1.25	1.75
N 242 black	50	50
Santocure	1	1.25

Yet another phenomenon was observed, described in Fig. 5. Samples of the reinforced rubber were pulled to 200% extension at 30 °C at a fixed speed, and the following conditions were imposed before testing:

a) Immediate release on attaining the extension of 200%
b) Maintenance of this extension for 5 sec and
c) for 60 sec
d) as c) with recovery at room temperature for 4 weeks.

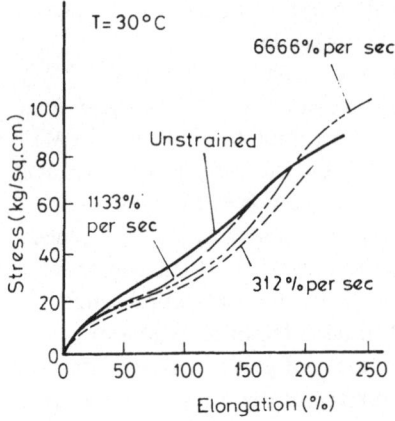

Fig. 2. Stress-elongation curves at various rates of elongation[75]. Compound 1403-1

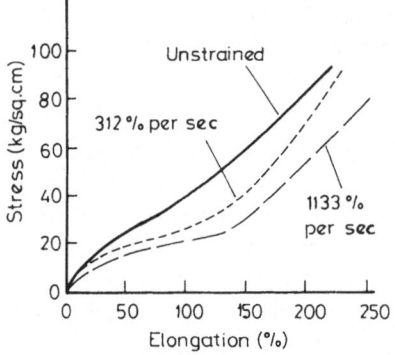

Fig. 3. Stress-elongation curves after prestrain to 200% at various speeds and at reduced temperature[75]. Compound 1503-5, T = 5 °C

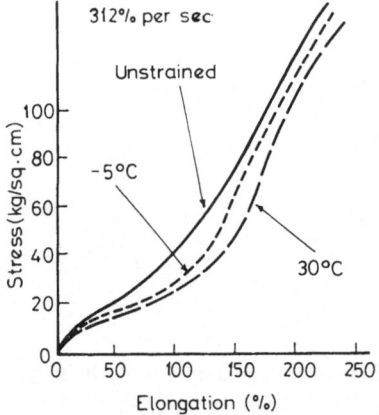

Fig. 4. Stress-elongation curves after prestrain to 200% at 312% per sec at various temperatures[75]. Compound 1403-5

The results indicate that the stress softening develops (probably asymptotically) with time during which stress is maintained, and that it is at least partly recoverable. Other workers state that stress softening is not fully recoverable even when the stressed samples are allowed to relax in a swelling solvent.

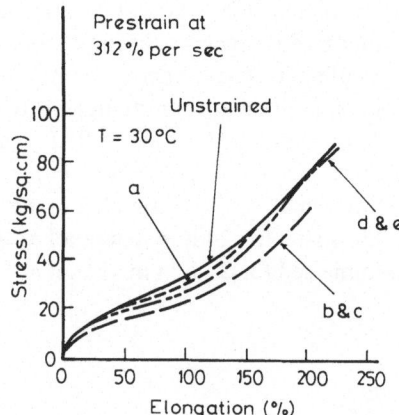

Fig. 5. Stress-elongation curves after prestrain to 200% at room temperature. See text

7. Stress Softening as a Result of Swelling by Solvents

Following the speculations raised by the above results, it was considered important to devise experiments which would make it possible to observe stress softening in vulcanizates as a result of swelling by solvents. This phenomenon had not been observed before. The experimental scheme was based on the observation that stress softening apparently recovers faster the higher is its temperature (see Fig. 5) and that presumably, at temperatures approaching T_g no recovery should take place. Rigbi[75] showed that at 0 °C, recovery is extremely slow even in compounds for which T_g is of the order of −40°. However, when the rubber is swollen in a solvent, T_g will be much reduced. Therefore, as low a temperature as conveniently obtainable was planned for a series of experiments, reported by Rigbi[74]. (It may incidentally be noted that stresses in the macromolecular network do develop on swelling in solvent, but as long as no external stress is applied, the meaning of the expression "stress softening" becomes debatable. Nevertheless, this semantic problem will not be discussed here).

The following precautions were taken in these experiments in order to minimize recovery:

(i) The temperature was reduced to the lowest possible level during the drying out of the solvent.

(ii) The swelling solvent was removed by means of immersion in a large excess of a low-viscosity non-solvent miscible with the swelling solvent. This process will be henceforth called deswelling.

(iii) A combination of both.

Strips 1/4″ wide, cut from the vulcanizates which contained no softening oils, were extracted by immersion in chloroform, and dried at room temperature. These were then subjected to various swelling and deswelling treatments as follows:

1. As received after extraction; control group.
2. Swollen in chloroform at room temperature; dried at room temperature under vacuum overnight.

3. Swollen in chloroform and deswollen in methanol, both at room temperature, followed by vacuum drying at room temperature for 1 h.
4. Swollen in chloroform at room temperature, deswollen in methanol at −8 °C, vacuum dried at room temperature for 1 h.
5. Swollen at −8 °C, dried at R. T. overnight.
6. Swollen at −8 °C, deswollen in methanol at R. T., vacuum dried for 1 h at R. T.
7. Swollen at −8 °C, deswollen in methanol at −8 °C, vacuum dried for 1 h at R. T.
8. Immersed for 24 h in methanol at R. T., vacuum dried for 1 h at R. T.
9. Immersed for 24 h in methanol at −8 °C, vacuum dried for 1 h at R. T.

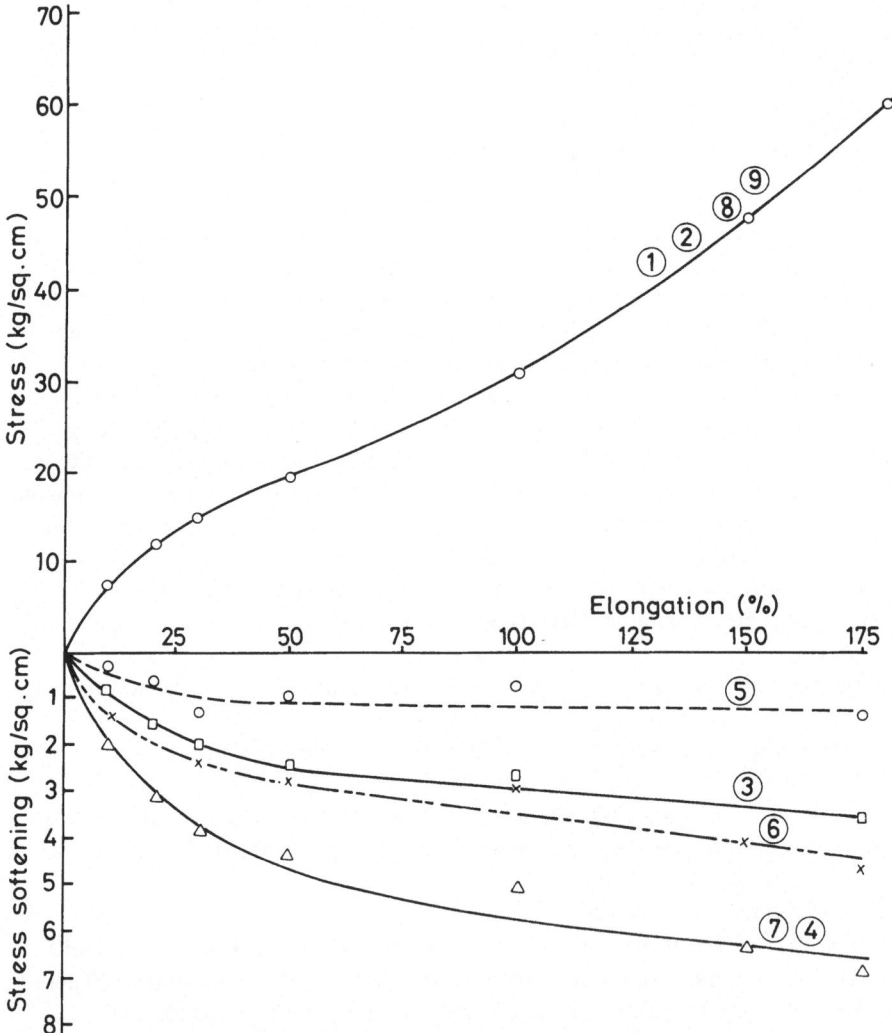

Fig. 6. Stress softening of Compound 1403-1 after swelling in chloroform. Numbers refer to deswelling treatments described in text[74]. Reprinted by permission of John Wiley & Sons, Inc.

Three strips (gauge length 10 cm) of each group were tested in an Instron tester at crosshead speed of 100 cm/min care being taken not to stress the samples during handling. Stress readings were compared for each elongation by averaging the three values obtained for each group. The results are given in Fig. 6. These show conclusively that:

a) Methanol alone causes no change in the stress-strain properties either at R. T. or at −8 °C.

b) Swelling in chloroform and drying at R. T. results in no apparent change in the stress-strain behaviour.

c) Deswelling in methanol at −8 °C followed by vacuum drying at R. T. for one hour results in the greatest stress softening in the series.

d) Deswelling in methanol at R. T. results in rather less stress-softening, although swelling in cold chloroform appears to soften the rubber more than swelling in R. T. chloroform. This is presumably due to the fact that at the instant the sample is immersed in the R. T. methanol, it is cold.

e) Even drying at R. T. after swelling at −8 °C results in some residual stress-softening.

8. Thermodynamics of Stress-Strain

Oono, Ikada and Todani[66] analysed the thermodynamics of black reinforced SBR on stretching, and showed that as the proportion of carbon black was increased from zero, a continuous increase in the energy contribution to the total stress developed. At low loadings, and an extension of 100%, the stress developed is due principally to a very considerable change in entropy which compensates for a fall in the internal energy. As loading is increased, more and more of the stress derives from changes in internal energy, which becomes responsible for the major part of the stress above loadings of 75 phr.

9. Interpellation

We must now summarize the information, and direct it towards answering the question raised in the introduction: what is it in carbon black that makes it behave so differently when compared to other fillers? To this end, the following tabulation will be helpful:

1. Fillers, with the exception of carbon blacks and a certain silicas and silicates do not generally increase the strength of elastomers. Particularly in non-polar and non-crystallizing polymers, the strength increases in very great measure as the result of the introduction of carbon black. Here, strength refers to tensile or tear strength or to abrasion resistance.

2. In parallel with this increase, certain undesirable properties which indicate viscoelastic behaviour become more marked as the proportion of carbon black is raised. These include stress-softening, compression set, hysteresis and heat build-up. Only comparatively insignificant increases become apparent, if at all, when non-reinforcing fillers are used.

3. Equal increases in modulus resulting from the use of (a) carbon black and (b) other fillers bring about (a) increased and (b) decreased abrasion resistance, respectively.

4. Immobile attachments between elastomer and filler are, in the writers opinion, generally deleterious to mechanical properties (e. g. brass as a filler).

5. Stress-softening in moderately loaded black vulcanizates increases with time and recovers (partly) with time. The first effect becomes more pronounced as temperature falls and the second effect more rapid in the presence of solvents and at higher temperatures. At temperatures near the glass transition, carbon black reinforces SBR very little indeed.

6. Some elastomer becomes attached to the carbon black during mixing, and can only be partially removed by solvent extraction at any temperature below a given value which is probably characteristic for a specific polymer and a specific carbon. Dr. E. Papirer has indicated in the course of discussions that attempts to remove attached elastomer by solvents boiling near the T_m value always result in residues greater than would have been expected from the extrapolated line. The reason for this is not clear, and may be due to induced crosslinking[85]. Grafted polymer may likewise be removed, but T_m is then considerably higher than for adsorbed solvents.

7. All elastomer can be removed from nonreinforcing fillers by any solvent at all temperatures. While the author cannot quote the literature for this statement, direct and indirect evidence for the complete separation of cured elastomer from fillers such as chalk, talk and clay, when swollen in solvents[7] make this statement more than probably true.

In the opinion of the writer, this characteristic behaviour of carbon black requires the assumption of an unusual attachment between the elastomer and the black. One such mechanism has been termed a "slippage mechanism" by Dannenberg[19] and other writers, but has not been well received by researchers in this field. One reason is, without doubt, that slippage is not related to a physical phenomenon capable of better description than that of a strand of spaghetti being pulled out of a mess of sphagetti in tomato sauce. Another reason is that the very term slippage does not indicate any recovery process. In the theory which follows (Sect. 14) the term slippage will be completely avoided and *saltation* used instead[3]. This theory, inspired by the Eyring theory[26] of viscous flow, was first formulated in 1968[77].

It does not follow that the mechanism described by the theory is the only mechanism involved in the behaviour of carbon black filler in rubber: the "geometric" or "hydrodynamic" effect resulting from the Guth-Gold factor properly applied and the particle size continue to play their important part; occluded rub-

---

3 **Saltare** – To leap. Hence *saltation*. The noun jump or *saltation* will be for a single step in the process, and terms such as "saltation process", "chain extension by saltation", "saltation equilibrium", etc. . . . will be used. The infinitive "to saltate" and the gerund "saltating chains" will also be used occasionally

ber will influence the overall behaviour in the same way as at present described; and aggregate interactions and aggregate breakdown will very considerable influence the picture. But most of these will also be characteristic of many nonreinforcing fillers at the same geometry. Their influence must be all superimposed on the phenomena resulting from saltation, this being unique in simultaneously explaining the points made above.

10. Behaviour of Filled Polymers Near Rupture

The phenomena observed in rubber as it nears rupture and subsequently have not been studied with the same attention as behaviour at lower stresses[4]. Considering a vulcanizate as a single uniform material, rupture has been studied principally from by means of the tensile test, the tear test in its various modification, abrasion testing and hysteresis effects.

a. *The Tensile Test.* Apart from its importance in the routine control of rubber manufacture, Smith[87, 88] has been able to extract considerable information from the tensile test by means of the "failure envelope". The failure envelope was also used by investigators such as Halpin[36], and others to reach interesting conclusions. Failure envelopes are obtained as the line joining the loci of points, reduced to a standard conditions, representing the stress at rupture and the corresponding elongation obtained at various temperatures, rates of stressing etc. As pointed out by Eirich[25], viscoelastic time-temperature superposition phenomena in the sense of Williams, Landel and Ferry[100] must be involved in fracture processes to an appreciable extent. The WLF correction for temperature notwithstanding, the data presented by Smith[88, 89] show that points supposedly distributed about a single curve in fact lie along distinct lines representing temperature and intersecting the envelope. This is particularly noticeable on the upper branch of the envelope but exists where the slopes of the lines are close to that of the lower branch of the envelope. It is therefore possible that failure envelopes are actually artifacts which represent real behaviour only approximately while hiding "fine structure" details.

In examining such envelopes, attention must be paid to the possible influence of filler on crystallization in the matrix. Eirich[25] quotes Studebaker's findings that crystallization is increased in the presence of fillers. This is probably the direct result of strain magnification (discussed in Sect. 5). In non-crystallizable polymers, of course, strain magnification causes no such effects, and envelopes for the two types of polymers are therefore not comparable.

Nevertheless, because data from tests at constant strain rate, constant strain and constant load yield substantially the same failure envelopes[35, 37], these studies lead to important conclusions. Assuming that a theory of strength and failure for gum

---

4   The attention of the author has been drawn to an interesting study by Bartenew [Plaste u. Kautschuk, *17*, 235 (1970)] in which the correlation of viscoelastic behaviour and rupture properties is attempted. This deals only with gum elastomers, but when it will be extended to reinforced vulcanizates, it may throw additional light on the subject of this review

vulcanizates exists, comparison of failure envelopes for gum and reinforced vulcanizates may indicate why carbon black acts to reinforce elastomers.

Unfortunately, the literature is quite poor in such studies, which are very time consuming. One important paper by Halpin and Bueche[37] presents failure curves for gum-SBR and vulcanizates filled with 15 and 30 phr HAF black. Their failure envelopes (Fig. 4, Ref.[35]) are not particularly revealing, except to show that for any reduced stress at break, the corresponding elongation at break is smaller the greater the proportion of black introduced into the mix over much of the range. However, the indications are that at high values of stress, as would be obtained at lower temperatures or higher speeds of straining, the curves tend to cross over.

Of greater interest is a comparison of the curves of time-to-break vs. reduced stress (Fig. 10, Ref.[37]) and time-to-break vs. extension at break for the gum and unfilled rubbers. (The second curve has been replotted in Fig. 7 of this review from the data in both Fig. 6, Ref.[35], and Fig. 1, Ref.[37]).

As pointed out by Halpin and Bueche, increasing the filler concentration, results in a rapid increase of strength at a constant time to break, t_b, and this is accompanied by an increase in the extension at break. Thus at $log(t_b/a_t) = 3$, the increase in strength is about 70% for a loading of 15 phr, but the elongation is about doubled. This increase in strength is approximately constant in the range of $log(t_b/a_t) = 0$ to 3, and represents the maximum attainable with 15 phr HAF. On the other hand, the curves for λ_b are roughly parallel and the improvement is greatest over a shorter range, $log(t_b/a_t) = -1$ to $+1$.

Bueche and White[15] observed the progress of rupture in a tensile specimen. Because the rate of crack propagation reached a maximum which was equal to the veloc-

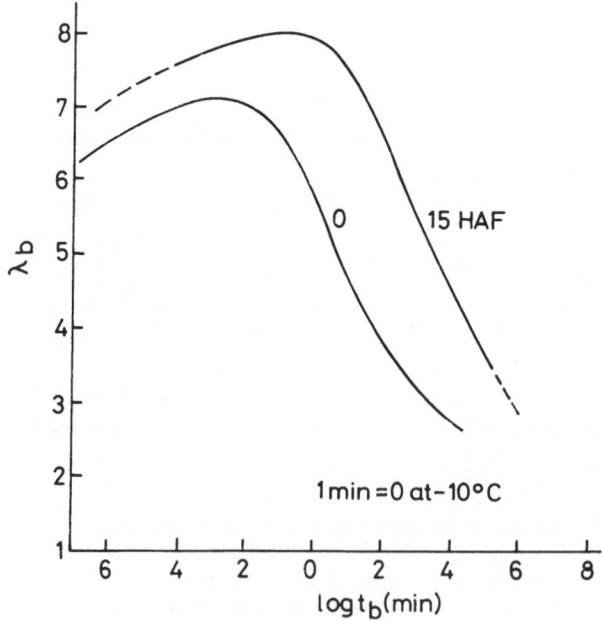

Fig. 7. Extension ratio at break for various time to break of gum and lightly loaded SBR. (after Halpin[35] and Halpin & Bueche[37])

ity of the stress wave calculated from modulus and density values for the elastomer in the glassy state, they considered rupture to be "brittle". Their observations were confirmed by Mason[60].

In Halpin[36], the molecular theory (of strength or fracture) of Bueche and Halpin is given in some detail. This theory results in specifications for criteria for rupture of a chain in terms of (a) a critical stress, (b) a critical deformation or (c) a critical value of stored energy. These are then related to the total time required to unfold a convoluted chain so that it effectively supports the load along a single length rather than a larger number of folds. Of the three criteria mentioned, it is easiest to discuss (a) because of its explicit form. This states that for chain rupture, $F_0/(\tau_1/N^2 \vartheta)^{1/2} \geqslant$ $\geqslant F_c$. Here F_c is the critical force necessary to break the chain, F_0 is the applied force, N is the number of monomer units forming the chain, τ_1 is the longest retardation time of the chain and ϑ is the time elapsed after application of the load.

It is now possible to compare a loaded vulcanizate to a gum compound. Gregory, Metherell and Smith[32] show that for loadings of SRF black, the relaxation rate increases with proportion of black and with strain. It also varies linearly with MR 100 (equilibrium modulus at 100% extension), and the stress relaxation rate at 100% increases with the fineness of the black, going from 2.1% per decade for N110 black to 4.1 for N762, 4.7 for N330 to 5.1 for N110. The relaxation rate bearing an inverse relationship to the retardation time τ_1, it follows that as carbon black loading increases (or the black at a given loading is made finer) application of a given force will result in rupture of a particular molecule after a shorter time.

It should be noted, however, that as the *overall average* relaxation rate of the filled vulcanizate increases, it must do so very much more steeply at or near the surface of the carbon black, since obviously, the relaxation characteristics of the matrix itself do not change. Therefore the crack tip, as it advances, will search the path of least resistance, this passing from black particle to particle, extending the path very greatly. (This is the development of a knotty tear). Hence the time under load of the molecules distant from the crack tip increases to raise the value of F_0, or σ_b.

At this stage, it is not possible to calculate the values of (τ_1/ϑ) for a loaded vulcanizate, since the Bueche-Halpin theory deals with a particular macromolecule, and must be extended to a distribution of molecular lengths in a matrix, and to the lengths between crosslinks and carbon black particles.

That the rupture path passes very close to the surface of the black particles, whose adhesion to the matrix is seriously weakened by the stress immediately proceeding rupture, has been demonstrated by Kamenski et al.[45].

Surfaces relating tensile strength of vulcanizates as a function of temperature and carbon black loading parameters, have been given by several authors[81], but throw little light on theory.

b. *Crack Propagation.* The propagation of cracks in stressed elastomer sheets has been considered by Rivlin and Thomas[73], Greensmith, Mullins and Thomas[31] and Knauss[47]. While the basis to the theory developed by Knauss and the choice of critical parameters, is different, his experimental results confirm Halpin and Bueches[37] prediction that rupture energy, as defined, should increase rapidly with rate of crack propagation.

The tearing or rupture energy measured by Rivlin and Thomas[73] on black mixes shows humps or maxima at characteristic rates of tearing, except at higher temperatures. This is the result of a viscoelastic process which was shown not to exist in gum rubbers.

An experimental study of considerable importance was published by Glucklich and Landel[30]. By means of accurately moulded tear tips of prescribed radii and suitably dimensioned trouser test pieces according to Rivlin and Thomas[73], they were able to show that the strain energy density concentrated at the crack tip was much higher than the strain energy density in a tensile test specimen at the point of rupture. The ratio of these densities depend upon the nature and concentration of the fillers, falling much more rapidly from a value of about 10 to about 1 for HAF than for MT. The minima in these curves correspond to maxima in the curve for the crack tip diameter formed by the test piece. Glucklich and Landel present curves for tearing energy at various tearing rates and temperatures. The evidence is that the maximum tearing energy is of the same order for both blacks, this occurring at a volume loading of 0.35 for MT and 0.25 HAF black. However, the intrinsic strength as defined by Lake, Lindley and Thomas[56] is higher for HAF by a factor of about 3.

c. *Abrasion.* The mechanism of abrasion has been studied amongst others by Shallamach[83] who has shown that it is a complicated process of repeated straining of small volumes at the surface resulting finally in tear and their separation from the bulk of the rubber. Hence, after a number of cycles of prestress, abrasion loss is the result of a high speed tensile process (Ecker[24]).

Kraus[50] has analyzed the effects of the addition of carbon black on abrasion and wear in his excellent review. He states the relation

$$\frac{A}{\mu} \sim \frac{C}{U_b}$$

A being the volume abraded, μ the coefficient of friction and U_b the energy input to break. Presumably, because A is so highly dependent on the test-method used, C is determined not only by "some unknown interactions of test conditions and properties of the sample being abraded" (Kraus[50], p. 226) but by the test method itself. Being that as it may, Payne and others[39, 40, 51] show that $U_b \propto (H_b \epsilon_b)^{1/2}$, H and ϵ being the hysteresis and elongation respectively, both at break. As μ is not very sensitive to carbon black additions, $A \propto (H_b \epsilon_b)^{-1/2}$. Although considerable information is available concerning hysteresis losses at small deformations, nothing has been found to relate H_b to carbon black loading. Nevertheless, if both measures of hysteresis are the result of the same mechanism, H_b should increase with loading and inversely with particle size. From the information in Fig. 7, it is seen that for any given time-to-break t_b, ϵ_b must increase with loading. The abraded volume must therefore fall with carbon black loading, as it does in fact.

11. Two-Phase Model of Matrix Behaviour

Blanchard and Parkinson[6] postulated two types of attachments between a rubber matrix and its carbon black filler: a "strong" type due to chemisorptive attachments and a "weaker" van der Waals type of bond. The former were considered to remain unbroken upon stressing the vulcanizate, but the latter suffered progressive rupture as the stress increased.

Blanchard developed the model in additional studies[4, 5] and explained certain behaviour as the result of the existence of two phases in the matrix – a hard phase and a soft phase. This model was also proposed by Mullins and Tobin[64]. As a result of stressing, increasing proportions of the hard phase become softened. The subsequent behaviour is then controlled by the model described by Takayanagi et al.[95], which was also used by Gent[28] and Blanchard[6].

The model can be used to describe many of the phenomena observed, but it is strictly a mechanical model which does not explain why these phenomena occur.

12. The Basic Model for the Behaviour of Carbon Black in Rubbers: Single Macromolecules Under Tension

In Fig. 8, let A be a carbon black aggregate, and M an elastomer macromolecule attached to a point F in space. M contacts A at a point P along the molecule and a point S on the surface of A. The character of F does not concern us at the moment, and at this stage of the discussion, A is fixed relative to F.

S is one of many points of similar chemical character on the surface. Its nature is of no interest at the moment except that it is such as to allow the adsorption of the elastomer molecule at P with a certain bonding energy E_a. In the vicinity of S there may be other identical points S', or others offering a bonding energy less than E_a. Equally, at given distances measured from P along M, there will be other points P' and P'' capable of bonding to S and S' in a similar way.

The chain is under no *average* tension. However, because of thermal movements of the chain or parts of it, the energy transmitted to P may at some moments rise above E_a by at least an excess E, when the chain will become momentarily detached from S and then reattach itself. Attachment at S' is possible but less likely, due to

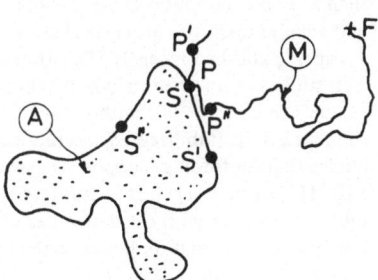

Fig. 8. Single elastomer macromolecule M fixed at point F adsorbed at P and S to carbon particle A

other attached chains in the neighbourhood, these creating a "cage effect" around the particular chain concerned. At a later instant, another such movement will take place, most likely but not necessarily in the reverse direction.

There is a certain difficulty in the consideration of zero average tension. While the most probable end-to-end distance of an unstrained molecular chain is $r = a\sqrt{2N/3}$, and the root-mean-square value for r in a Gaussian distribution is $\langle r \rangle = a\sqrt{N}$ (Treloar[93], p. 56), a tension in the chain exists for all values of r (loc. cit., p. 62), including $\langle r \rangle$, other than zero. We do not resolve the difficulty here, but for our purpose assume that the chain is under no tension different from that which would be found in a free chain of the same length.

These detachments and reattachments, which will be called saltations, occur sporadically, but P and S represent statistically mean points. The value of E is not at present known, nor is it known whether E and E_a are related. E is the energy required to activate the process of saltation, or in brief, the activation energy of saltation.

It is now necessary to speculate on the preferred point of attachment on the carbon black surface. If the chain emanating from F is very long, would it preferably attach to S or to S''? In what follows, a^5 is the length of the monomer unit, b is the straight-line distance FP, and Na the distance FP measured along the chain. Let it be assumed that the length of the whole chain is much greater than b, where now $b = $ FS''. The chain will attach to any point such as S'' which it approaches. Following the instant this connection is made, one of several processes may now occur:

a) If $N^1 a > b^2/a$ (this being the effective chain length between F and S'') then the chain will begin a series of mostly backward jumps, until $Na = b^2/a$. Alternatively, and more likely, due to thermal movements in the chain, a monomer unit may come sufficiently close to the surface at S' or S'' to become attached to it, and then the chain may become doubly connected, as visualized by Silberberg[84].

b) If $N^1 a = Na = b^2/a$, the effective chain length will not change until a situation as described in the second alternative of (a) occurs.

c) If $N^1 a < b^2/a$, the chain will be under tension, and jumps will relieve this until the system reaches a situation as in (b).

It can be seen that no matter how the macromolecule is initially disposed in relation to the particle of the black, it will ultimately attach itself (on the average) to the closest high-energy point to F, and its most probable length will be $Na = b^2/a$.

---

5   It is assumed here that in every monomer unit there is one point such as P, and that the distance of a single jump is a. However, in certain polymers there may be two identical points such as P in one monomer unit, while in random copolymers PP' may not be constant and will depend upon the order of chain growth and the presence of cis- and trans-configurations. A paper published recently [K. Kozlowski, Acta Polymerica, *30*, 547 (1979)] deals with ESP studies of uncured carbon black/natural rubber mixes, their solvent extracts, and the residues therefrom. The author found it possible to identify a very narrow spectral line with rubber radicals stabilized by interaction with active sites on the carbon black surface. He concludes that his findings support Meissner's theory [B. Meissner, Rubber Chem. Technol., *48*, 810 (1975)] that "each structural unit (of the rubber molecule, Z. R.) has the same probability of reaction with the active site of a carbon black particle and can form with it only one bond". The relation between the evidence adduced and Meissner's theory is not, however, clear to the writer

The hypothesis whereby the molecules saltate so that a point of attachment moves backwards and forwards is based upon Eyring's[26] theory of liquid flow, and like in his development, each jump requires the transfer to and from the particular portion of the molecule of sufficient energy to surmount an energy barrier of E kilocalories. The frequency of jumps of one monomer unit in either direction, forward or backward, about the most probable point of attachment is

$$\omega_0 = Z\ exp\ \{E/\kappa T\} \tag{1}$$

We now assume that the point F is moved a distance $b\,\epsilon$ away from S in the direction SF, as a result of which a force f develops in the chain. According to Treloar[93] this will be

$$f = \frac{3\,\kappa\,T}{N_0 a^2}\,b\,\epsilon. \tag{2}$$

Following Eyring[26], the rate of forward jumps will then increase and the rate of backward jumps decrease, giving a net forward rate, ω,

$$\omega_{+f} = Z\ exp\left(\frac{-E}{\kappa T} + \frac{3\,b\,\epsilon}{Na}\right) \tag{3}$$

$$\omega_{-f} = Z\ exp\left(\frac{-E}{\kappa T} - \frac{3\,b\,\epsilon}{Na}\right) \tag{4}$$

$$\omega = \omega_{+f} - \omega_{-f} = Z\ exp\left(\frac{-E}{\kappa T}\right) \times \left\{ exp\left(\frac{3b\,\epsilon}{Na}\right) - exp\left(\frac{3b\,\epsilon}{Na}\right)\right\}$$

$$= 2\,Z\ exp\left(\frac{-E}{\kappa T}\right) \sinh \frac{3b\,\epsilon}{Na} \tag{5}$$

which may be approximately written as

$$\omega = Y\left[\frac{3b\,\epsilon}{Na} + \frac{1}{6}\left(\frac{3b\,\epsilon}{Na}\right)^3\right]$$

$$\tag{5a}$$

where $Y = 2\,Z\ exp\,(-E/\kappa T)$

But the frequency is the rate at which the number of monomer units in the chain grows, that is $\omega = dN/dt$, and this equation may be integrated to give

$$Y\vartheta = \frac{N^2 - N_0^2}{2\,\rho^2} - \frac{1}{12}\ ln\ \frac{\rho^2 + 6\,N^2}{\rho^2 + 6\,N_0^2} \tag{6}$$

where N = instantaneous value of chain length
$\quad\quad\quad\ N_0$ = initial chain length
$\quad\quad\quad\ \rho$ = $3b\,\epsilon/a = 3\,\epsilon\sqrt{N_0}$
$\quad\quad\quad\ \vartheta$ = elapsed time since application of movement of F.

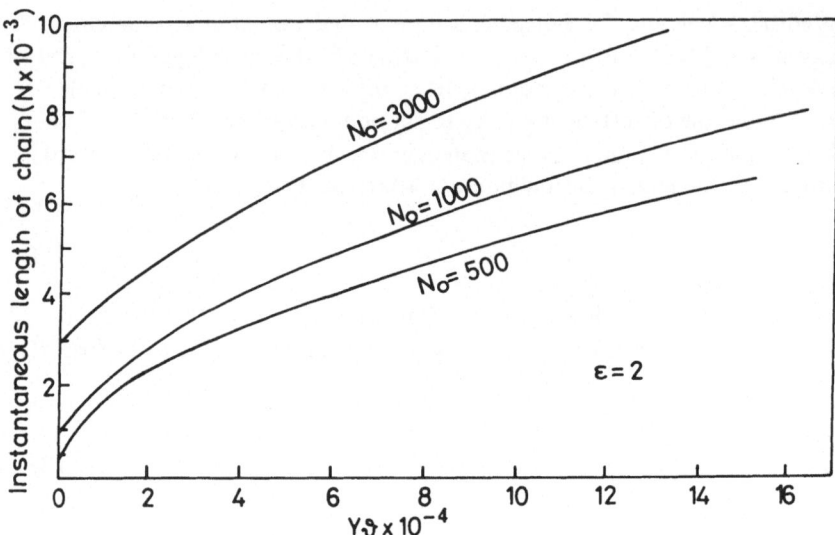

Fig. 9. Instantaneous length of chain as a function of time elapsed for various initial lengths N_0

Note that for very large values of N, the hyperbolic sine may be approximated by a single term, which reduces the model to a "dash-pot".

A plot of the function is given in Figs. 9 and 10. Figure 9 shows the growth of the chain in monomer units as a function of the time elapsed following the move-

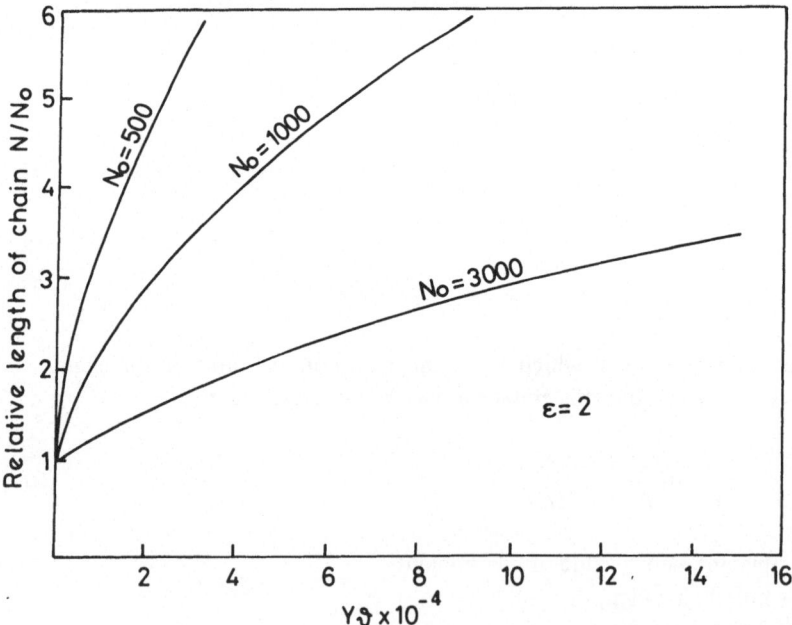

Fig. 10. Relative length of chain as a function of time elapsed for various initial lengths N_0[77)]

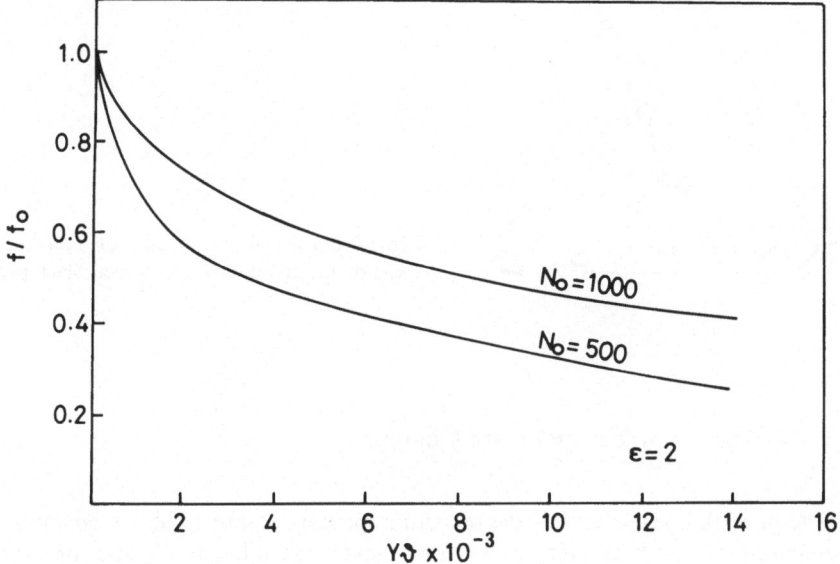

Fig. 11. Relaxation of single elastomeric molecules in contact with carbon black[77]

ment of F. Figure 10 shows the same function but as a multiple of the original chain length.

These curves are slightly different from those appearing in the original paper by Rigbi[77] in which the dependence of ρ on N_0 was neglected.

The logarithmic term in the expression developed (6) becomes rapidly insignificant in comparison to the first member as the ratio N/N_0 increases.

As the chain length increases, the force on the chain decreases according to Eq. (2). The way this develops has been plotted in Fig. 11, in which the ordinate indicates the proportional fall of the force acting along the macromolecule as a function of time.

13. A Single Chain Fixed at Two Points Contacting a Carbon Black Aggregate

Figure 12 shows a carbon black particle in contact with an elastomer molecule of given length between two fixed points. Considerations of equilibrium, assuming that the force along the chain is constant due to saltation, show that on the average, the lengths of the portions of the chain will be in propotion to the distances F_1S, and F_2S, and *not* according to the square root. A doubly attached chain will therefore behave in different fashion to Figs. 9 and 10.

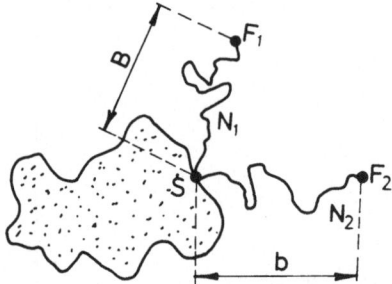

Fig. 12. Division of macromolecule attached to two fixed points and contacting carbon black particle

14. The Effect of Temperature Changes

The temperature sensitivity of the relaxation process described above operates through the factor Y. In fact, for a chain of given initial length N_0 and for a given step-movement of point F, we will obtain the same values of N provided that

$$Y_1 \vartheta_1 = Y_2 \vartheta_2 \tag{7}$$

or $\vartheta_1 \exp\left(-E/\kappa T_1\right) = \vartheta_2 \exp\left(-E/\kappa T_2\right)$

or $\vartheta_2/\vartheta_1 = \exp\left[-\dfrac{E}{\kappa}\left(\dfrac{1}{T_1} - \dfrac{1}{T_2}\right)\right]$

from which $\dfrac{1}{T_2} - \dfrac{1}{T_1} = \dfrac{\kappa}{E} \ln \dfrac{\vartheta_2}{\vartheta_1}$ \hfill (8)

These equations are shift equations for the effect of temperature and time. They may be interpreted in the following way: knowing the relaxation behaviour of the chain at one temperature, we are able to obtain the same degree of relaxation at a higher temperature if we allow the chain to relax for a shorter time specified by Eq. (8).

Of greater interest is the result obtained by differentiating Eq. (5), to obtain

$$\frac{d\omega}{dT} = \frac{E}{\kappa} \cdot \frac{\omega}{T^2} \tag{9}$$

Since ω has the larger value for shorter chains, the temperature dependence of the rate of chain growth increases as the chain becomes shorter.

15. Redistribution of Stress Between Two Chains

If two chains link a common point and a single aggregate of carbon black, and if these are so disposed that the straight lines joining the common point F and the points of contact with the black are almost collinear, the forces acting along the chains are universely proportional to the number of monomer segments in each. Hence if W is the total load, and W_1 and W_2 are the instantaneous forces in each chain, respectively, we shall have

$$W_1 = \frac{WN_2}{N_1 + N_2} \quad W_2 = \frac{WN_1}{N_1 + N_2} \tag{10}$$

Assuming that we are dealing with periods of time long enough for the second term in Eq. (6) to be negligible, i. e. that $Y\vartheta = \dfrac{N^2 - N_0^2}{2\rho}$ holds, we have

$$\frac{W_1}{W} = \frac{N_2}{N_1 + N_2} = \left[1 + \left(\frac{N_{10}^2 + 6Y\vartheta\epsilon\sqrt{N_{10}}}{N_{20}^2 + 6Y\vartheta\epsilon\sqrt{N_{20}}} \right)^{1/2} \right]^{-1} \tag{11}$$

Where N_{10} and N_{20} are the initial values of N_1 and N_0 respectively, and a similar expression may be written down for W_2/W. Figure 13 has been plotted to show how these values change with time for $N_{10} = 500$ and $N_{20} = 3000$. It is seen that as the

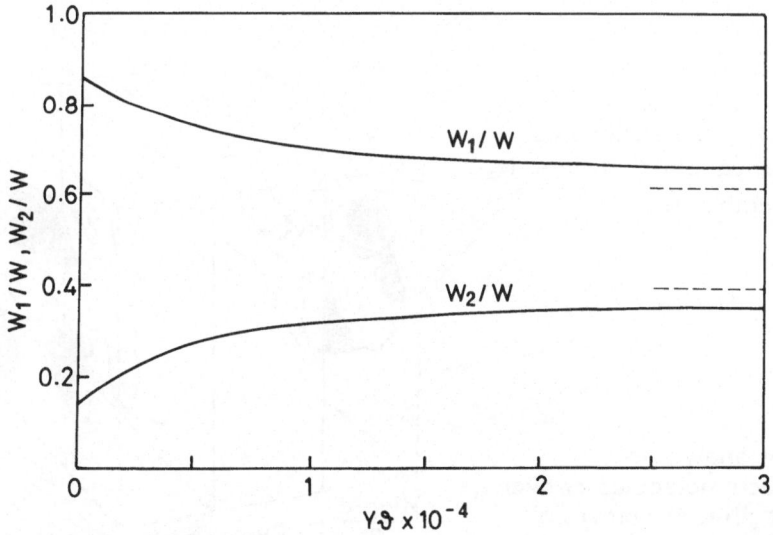

Fig. 13. Redistribution of load between two chains.
$N_{10} = 500 \qquad N_{20} = 3000 \qquad \epsilon = 2$
$W_{10}/W = 0.8571 \qquad W_{20}/W = 0.1429$
$W_{1f}/W = 0.610 \qquad W_{2f}/W = 0.390$

process proceeds with time, the load is progressively removed from the shorter chain and transferred to the longer chain, in other words, that a process of redistribution of load between the two chains is involved.

16. The Filled Vulcanizate

We now wish to extend the considerations above dealing with several different structures extending simultaneously within the same space. Thus, for example, a macromolecule attached to carbon black at one point along its length may be cross-linked to other molecules at one or more points on either or both sides of the black particle, while another may pass fairly close to the black without contacting it and be equally cross-linked at the distant points. In order to be able to develop the mathematical relationships required for the development of this theory, it is necessary to represent the state of affairs by means of a highly idealized model. However, the model proposed here is believed to contain all the elements of a uniformly filled and cross-linked elastomeric compound. The possible existence of crystallization and inter- and intra-molecular forces other than main-chain carbon to carbon links, cross links and bonds between rubber and carbon black as hypothesized earlier, is specifically neglected.

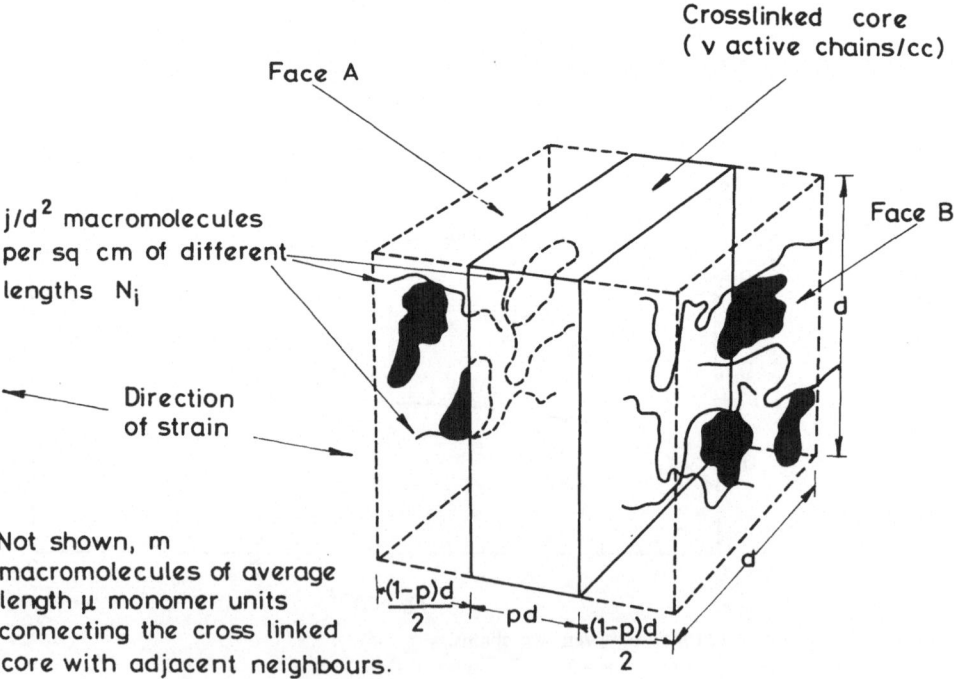

Fig. 14. Unit cell of model of vulcanizate

A diagrammatic representation of the features now to be described is given in Fig. 14. The vulcanizate is divided into a series of unit cells in each of which are found (*i*) a central laminar cross-linked core which contains no carbon black and which extends over the whole width and height of the cell: (*ii*) a number of carbon black particles which are located at the faces normal to the direction of strain, but which are otherwise randomly distributed; (*iii*) a number of macromolecules emanating from and bound to the central core, and attached to the carbon black by means of the mobile attachments previously described: and (*iv*) a number of macromolecules extending out of a core into the core of the unit cells adjacent to it in the direction of the stress. The dimensions of the unit cell are d in each direction and the width of the core is pd ($p \leqslant 1$). The molecules above are j in number and their lengths are N_i ($i = 1 \ldots j$), while those in iv are assumed to be m in number and of average length μ monomer units each of length a. The number of active chains per unit volume in the cross-linked core is ν.

This model of the cell may be assumed to duplicate that part of the behaviour of a real vulcanizate only in the direction of the unique normal stress. However, a large number of such cells, placed randomly with the carbon black particles acting as contact points between the cells, is presumed to follow the identical behaviour of a vulcanizate in all directions. The model therefore loses no generality in being obviously anisotropic. Because, in an actual vulcanizate, the core extends through the whole volume, we assume that the "end-to end" lengths of the attached macromolecules are not related to the dimensions of the basic cell.

A statically identical picture of the model is given in Fig. 15. Here, R and S are rigid surfaces, B is a black particle and H is a crosslinked mass of rubber, of crosslink density ν and thickness pd. H is connected to B through a saltating mechanism as described above by means of j chains of N_i molecular units ($i = 1 \ldots j$) and to S by m chains of μ molecular units. B is connected rigidly to S.

Assume S is now given a deflection δ. Forces f_i will act through the slipping molecules to give a total force

$$F = mf + \sum_1^j f_i$$

acting on H which will extend through a distance δ_H. The extension of the j and m molecules will therefore be $\delta - \delta_H$.

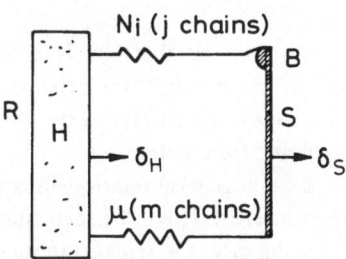

Fig. 15. Statically equivalent model of Fig. 14

Now the modulus of H will be $\nu \kappa T$, the stress acting upon is F/d^2, so that

$$\delta_H = pd \cdot \frac{F}{d^2} \cdot \frac{1}{\nu \kappa T} = \frac{F_p}{\nu \kappa T d} \, .$$

$$f_i = \frac{3 \kappa T}{a^2} \cdot \frac{\delta - \delta_H}{N_i}; \quad f = \frac{3 \kappa T}{a^2} \cdot \frac{\delta - \delta_H}{\mu}$$

$$F = \frac{3 \kappa T}{a^2} (\delta - \delta_H) \left(\frac{m}{\mu} + \Sigma \frac{1}{N_i} \right)$$

$$= \frac{3 \kappa T}{a^2} \left(\delta - \frac{F_p}{\nu \kappa T d} \right) \left(\frac{m}{\mu} + \Sigma \frac{1}{N_i} \right)$$

$$F \left[1 + \frac{3p}{\nu d a^2} \left(\frac{m}{\mu} + \Sigma \frac{1}{N_i} \right) \right] = \frac{3 \kappa T \delta}{a^2} \left(\frac{m}{\mu} + \Sigma \frac{N}{N_i} \right)$$

In deriving this relationship, a Hookean form of the stress-sextension law is assumed, rather than that derived by Kuhn[55], Wall[98] and Flory and Rehner[27], in order to simplify the mathematics. This will obviously change the details of the picture of internal and external deformations, particularly at high elongations, but the simplification is not thought to affect the ultimate conclusions seriously.

Noting that the modulus of the whole model is given by $F/\delta d$, we obtain the following expression for the reduced relaxation modulus:

$$\overline{M} = \frac{\text{relaxation modulus}}{\nu \kappa T} = \frac{\dfrac{m}{\mu} + \Sigma \dfrac{1}{N_i}}{p \left(\dfrac{m}{\mu} + \Sigma \dfrac{1}{N_i} \right) + \dfrac{\nu d a^2}{3}} \tag{12}$$

It must be noted that the individual values of N_i are both time and strain dependent, as has been shown above. In the present situation, this dependence is expressed by

$$(N_i)^2 = (N_i)_0^2 + \frac{a d^2 Y \vartheta \epsilon}{p \left(\dfrac{m}{\mu} + \Sigma \dfrac{1}{N_i} \right) + \dfrac{\nu d a^2}{3}} \tag{13}$$

ϵ being the overall strain. It is seen that it is not possible to rewrite this into an explicit expression for N_i; the calculation of N_i for selected values of the other parameters may be carried out by iteration, and the procedure is ideally suited for programming for a computer.

Several general relationships which may be derived from the above equations are of interest. It will be seen that the values of N_i increase with time, but at a diminishing rate; the relaxation modulus should therefore decrease until it reaches a constant level. Furthermore, if p vanishes, that is, if all the chains in the unit cell are connected to nodal points, the relaxation modulus will range from a maximum of $(3\kappa T/da^2) \cdot (m/\mu + \Sigma\, 1/N_i)$ to a minimum value of $3\kappa Tm/\mu da^2$ at infinite time. Again, if $j = 0$, we have the case of a gum rubber, for which the modulus is $3mv\kappa T/(3pm + \mu vda^2)$ which is independent of time as it should be. Both the value of the modulus immediately after the application of the strain and its final value are directly proportional to the temperature. However because N_i is a function of $Y = Z\, e^{-E/\kappa T}$, the rate at which the modulus changes is not a simple function of the temperature: all that can be said by inspection is that, as the temperature rises, the modulus will reach its final value earlier, as observed in filled rubbers.

The relaxation modulus is also a function of the strain, and according to the above it is expected that the model should be capable of treatment by means of a temperature-time shift as well as a strain-time shift. Thus, it is the product $\epsilon\,\vartheta\,e^{-E/\kappa T}$ which defines N_i, and the value of the ratio of the modulus to the absolute temperature should be constant for a constant value of the triple poduct. For any fixed temperature, the modulus at a time ϑ_1 after application of a strain ϵ_2, should equal the modulus at a time ϑ_2 after application of a strain ϵ_2 provided

$$\vartheta_1 \epsilon_1 = \vartheta_2 \epsilon_2 \tag{14}$$

This is obviously true for $\vartheta = 0$ and $\vartheta = \infty$.

When considering a constant initial strain, similar operations will show that for $M_1/T_1 = M_2/T_2$.

$$\frac{\vartheta_1}{\vartheta_2} = exp\left[-\frac{E}{\kappa}\left(\frac{1}{T_2} - \frac{1}{T_1}\right)\right]$$

or

$$\text{Shift factor} = ln\,\frac{\vartheta_1}{\vartheta_2} = \frac{E}{\kappa}\left(\frac{1}{T_1} - \frac{1}{T_2}\right)$$

This has been proven previously for the single chain model and is the same shift factor as used by Bueche[14] to explain the temperature sensitivity of filled stocks. Bueche, however, does not describe the mechanism of the system for which he proposes the shift.

It is also of interest to note that the Eqs. (12) and (13) may be combined into the form

$$Z\left[1 - \frac{pM}{v\kappa T}\right]\frac{3\epsilon d}{a}\,exp\left(-\frac{E}{\kappa T}\right) = \frac{N_i^2 - (N_i)_0^2}{\vartheta} \tag{15}$$

Since the quantities contained in the left side of the equation are constants under the conditions of any experiment, it necessarily follows that the change of the squares of the chain lengths for each and every saltating chain is constant during any interval under strain.

As stated, the relationships derived are not easily handled by means of algebraic operations, but they are in a form suitable for solution by iteration on a modern computer, giving plots for relaxation modulus, the increase of chain length and the redistribution of load between the chains. The values of the parameters were selected in the following way in order to simplify the program: the maximum and minimum values of chain lengths, $(N_1)_0$ and $(N_j)_0$ were first assumed, and the range between them divided into $(j - 1)$ intervals giving a geometric series of N_j chain lengths. Obviously, such a distribution is not necessarily similar to that expected in a real vulcanizate, but in the absence of any measured distribution, this is sufficiently realistic to give a qualitative description of the physical processes proceeding in a vulcanizate.

The values of the parameters selected for the computation of the relaxation curves shown in Fig. 16 were as follows:

ν = crosslink density of core = 10^{18} per cc

$(N_1)_0$ = initial chain length of shortest macromolecule connecting core with black = 500 monomer units

$(N_j)_0$ = initial length of longest macromolecule connecting core with black = 5×10^4 monomer units

j = number of chains connecting core with black = 6

μ = chain length of macromolecules connecting cores 10^4 and 10^5 monomer units

m = number of above = 2 and 6

d = 6×10^{-7} cm

p = proportion of unit cube filled by core = 0.2 and 0.5

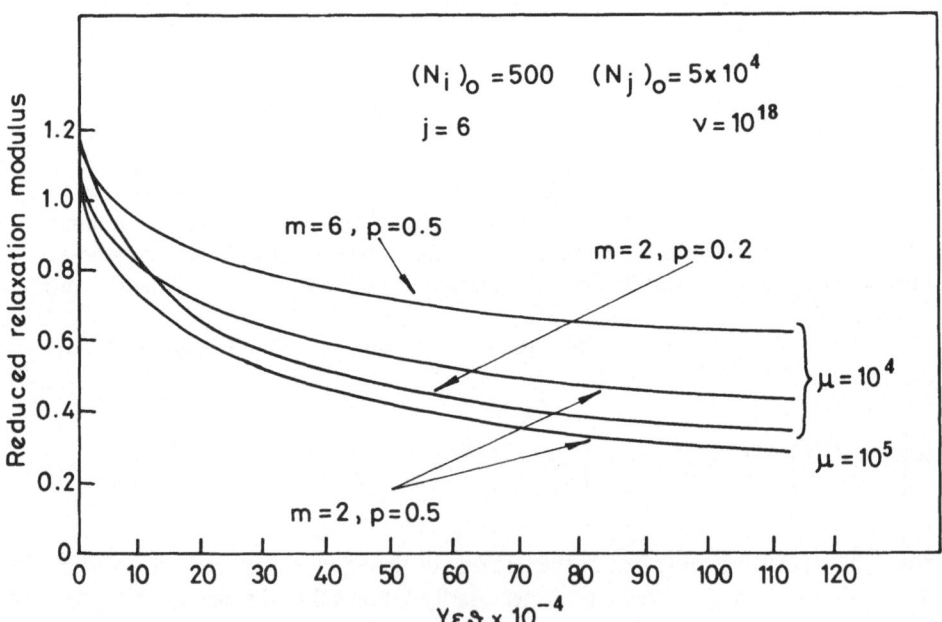

Fig. 16. Relaxation curves for model vulcanizate. Parameters as shown

Other sets of parameters were proposed; however, the above are qualitatively descriptive of all those investigated. It will be seen that decreasing the value p causes a change in the initial modulus, the nature of which (increasing or decreasing) depends upon the relative values of v, j and μ, and m and N_i. In the above case, the zero-time modulus increases as p decreases in value, but the relaxation is faster, since the non-relaxing core is diminished in importance, and the curves cross over to give a lower equilibrium modulus with $p = 0.2$ than with $p = 0.5$. It must be stressed that the ratio of the equilibrium moduli is determined by the relative values of the parameters, and the curves need not always cross.

When comparing the relaxation curves for $\mu = 10^5$ with those for $\mu = 10^4$, we see a higher modulus for the latter as would be expected. The same situation arises for a comparison of $m = 2$ with $m = 6$.

In the calculation of the above results, it was assumed that the dimensions of the elementary cube are 6×10^{-7} cm, that is, approximately seven times longer than the monomer unit (8.7×10^{-8} cm). At this point in the development of the theory, this appears to be somewhat too small to be related to any physical parameter of typical rubber compounds. The "average" end-to-end distance of an elastomer chain of 500 units is 22.4, that of 5×10^4 units is 224; these are significantly greater than the size of the unit cube.

These curves, when replotted on logarithmic paper, do not give straight lines, indicating that the model cannot approximate the type of behaviour described by Cotton an Boonstra[17]. Nor is the approximation by one relaxation time, as in a simple Maxwell model, at all satisfactory, However, the function may, of course, be converted into a Prony[6] series and the model may therefore behave in relaxation as a set of Maxwell models.

In a second series of calculations, parameters were chosen to be more realistic, as given hereunder:

$(N_1)_0$	= 500 monomer units (m.u.)	m	= 50
$(N_{48})_0$	= 50,000 m.u.	d	= $3 \cdot 10^{-5}$ cm
v	= 10^{18} per cc	μ	= 2500 m.u.
p	= 0.85	a	= $8.7 \cdot 10^{-8}$ cm

The response to an applied strain of 200% ($\epsilon = 2$) was then studied. As shown, the logarithmic term of Eq. (6) may be neglected for large N, and the response of the model then becomes equivalent to that of a generalized Maxwell body (Weichert body) with 48 discrete relaxation times, in parallel with a spring.

In this specific model, approximately the same number of chains connect the cores directly ("fixed length chains") as they do through carbon black agglomerates, and at zero time, $\Sigma (1/N_i) = 2.12 \cdot 10^{-2}$ and $m/\mu = 0.02$. Initially, therefore, the load is more or less equally divided between the two groups of molecules. In a relaxation experiment, when $Y\vartheta$ reaches a value of $60 \cdot 10^{20}$, $\Sigma (1/N_i) = 0.833 \cdot 10^{-2}$, and the chains of constant length can therefore be calculated as carrying about 71%

---

6 Prony series is the name given to a series of exponential terms, usually with the variable in the exponent negative and diminishing in absolute value. Reference is often made to R. Prony's paper in J. Ecole Polytechnique of 1795, but the derivation of the series that bears his name from his original paper is doubtful. See note by Z. Rigbi, Bull. Soc. Rheol., *23* (1), 1980, 14

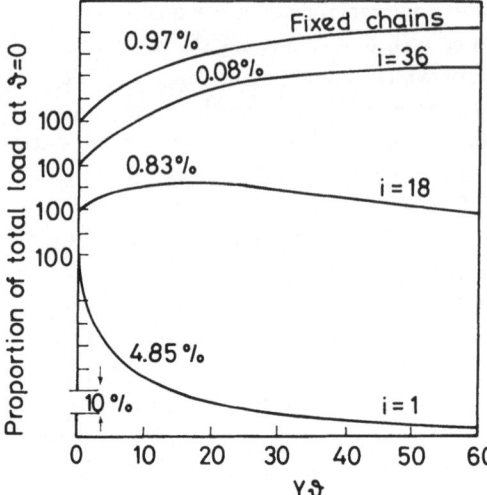

Fig. 17. Variation of load on some molecules in model of vulcanizate[76]

of the load. This redistribution of load has already been referred to in the case of the more primitive model ($j = 2$), and here too the shortest molecules grow most rapidly, shedding increasing proportions of the loads upon them to the other molecules. The variation of the load with time for the shortest molecule and two other molecules is shown in Fig. 17 and it is again seen that the load borne by certain molecules first increases and then begins to fall. The total relaxation of this model at an elongation of 200% is given in Fig. 18, and the variation of the lengths of the molecules with dimensionless time $Y\vartheta$ in Fig. 19.

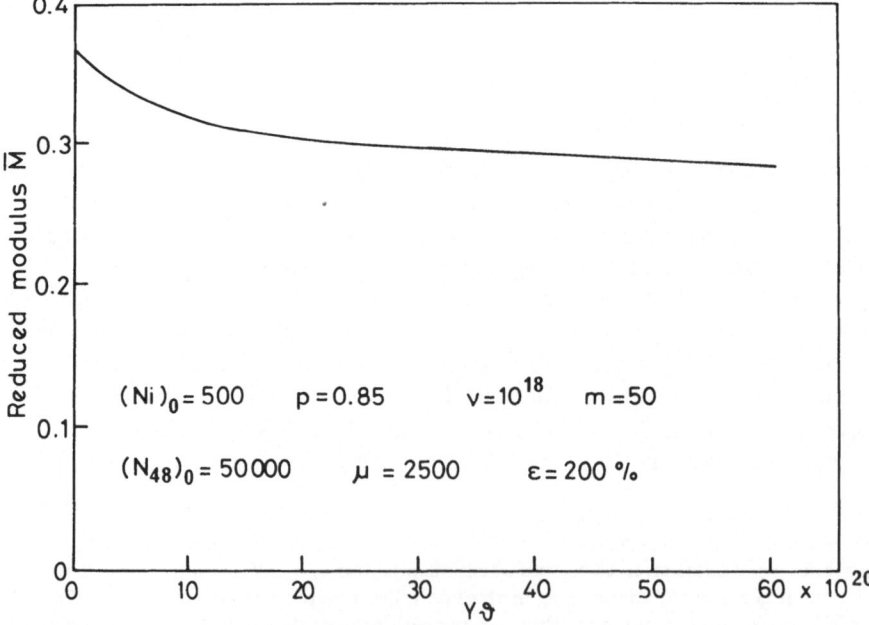

Fig. 18. Relaxation of model CB vulcanizate[76]

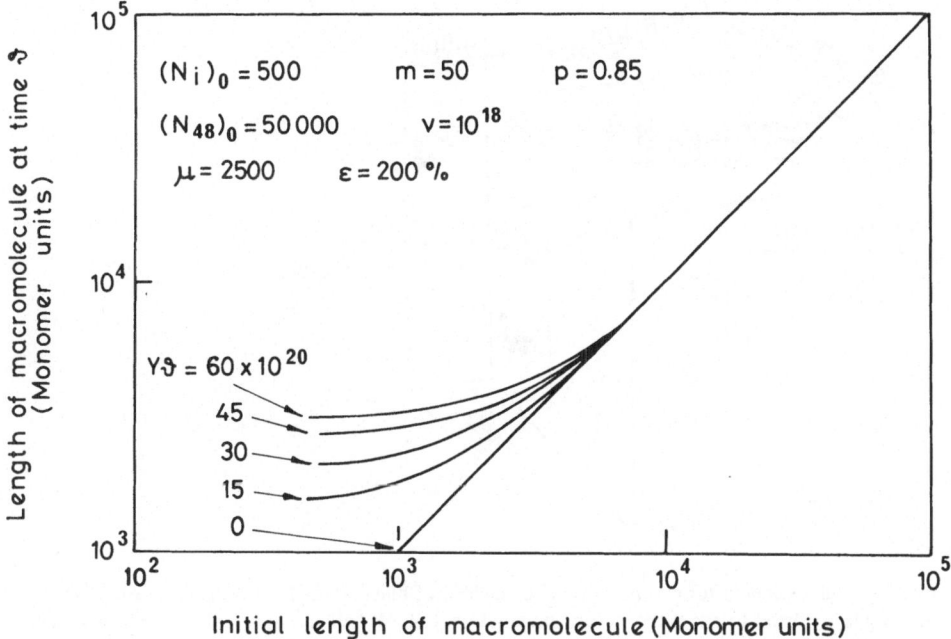

$(N_i)_0 = 500$ $m = 50$ $p = 0.85$

$(N_{48})_0 = 50\,000$ $v = 10^{18}$

$\mu = 2500$ $\epsilon = 200$ %

$Y\vartheta = 60 \times 10^{20}$

45

30

15

0

Initial length of macromolecule (Monomer units)

Fig. 19. Change of length of macromolecules in model of vulcanizate as a function of initial length and dimensionless time[76]

We are now in a position to translate the parameters of the second series in terms of the elementary cube. For a 40 phr loading of a typical black of 2.5 μm particle diameter (N-220 black), our model will consist of an elementary cube 30 μm high with cores occupying a width of 25.5 μm separated from nearest neighbours by spaces 4.5 μm wide. In these spaces, the carbon black particles are distributed. There are 11 of these per cell face, or about $5 \cdot 10^{14}$ particles per cc. The cores are attached to the black by 48 saltating molecules (i.e. about $17.7 \cdot 10^{14}$ saltating molecules per cc) of assorted lenghts between 500 and 50,000 units most of the molecules in the lower range, and interconnected by 50 molecules of length 2500 units per cell. Any section through a cell between the core and the black particles would cut $5.3 \cdot 10^{10}$ saltating molecules per sq.cm and $5.5 \cdot 10^{10}$ fixed molecules. Figure 20 shows the model with the relevant dimensions and other figures noted upon it. The calculated attachment density is about $25 \cdot 10^{12}$ attachments per square centimeter of black, much smaller than that obtained by Rehner[80] from other considerations.

The relaxation curve of Fig. 18, although it refers to a model with 48 discrete relaxation times, can be very closely approximated by a Prony series collocated at $Y\vartheta = 1, 10$ and $50 \cdot 10^{20}$ to give

$$\overline{M} = 0.2714 + 0.100675\, e^{-x} + 0.051\, e^{-x/10} + 0.0434\, e^{-x/50}$$

where $x = Y\vartheta \cdot 10^{-20}$

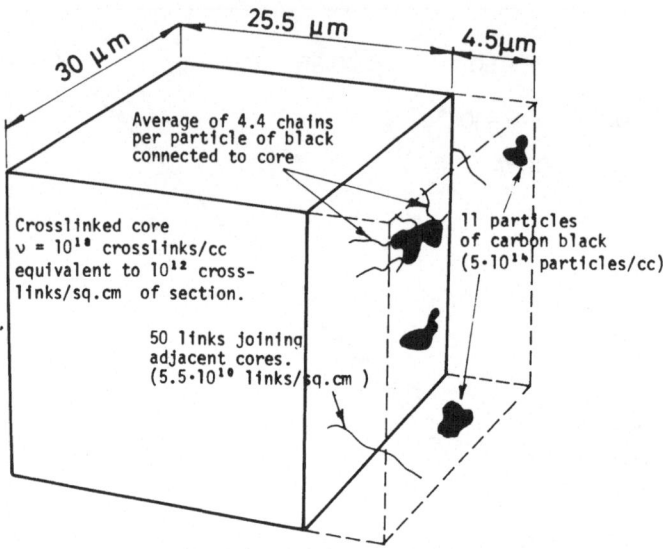

Fig. 20. Model selected for examination if behaviour of black loaded vulcanizate. (Note that drawing is distorted to show all relevant parameters. In fact, half the particles and saltating chains are on the invisible face)

Relaxation curves of loaded vulcanizates with similar crosslink densities, when cured with stable systems such as sulfurless TMTDS or with dicumyl peroxide, indicate losses of modulus of the order of 5% over periods of 10^7 sec at room temperature[16]. Comparison with Fig. 18 indicates that

$$10^7 \, Y = 4 \cdot 10^{20}$$

and therefore $Y = 4 \cdot 10^{13}$

As [see Eq. (5a)] this is half the saltation rate over the potential barrier at zero stress, it follows that the latter is of the order of 10^{14} per sec. This is rather greater than the value of $\kappa T/h$ (6×10^{12} per sec at 27 °C) which is obtained from Eyring's expression. However, considering the crude way in which the comparisons were made, the indications are that the proposed mechanism bears a close relationship to the real processes controlling stress-softening and, therefore, reinforcement.

17. Creep

If similar considerations are applied to creep under a total load $F \, (= \sigma d^2)$, the rate of change of chain length of individual chains will be seen to depend upon the relative lengths of the other chains emanating from the core and supporting the load; thus,

$$\frac{dN_i}{dt} = \frac{\sigma d^2}{2\kappa T} \cdot \frac{1}{N_i\,(m/\mu + \Sigma 1/N_n)} \cdot \qquad n = 1,2 \ldots j \tag{16}$$

Now, the total deformation will be

$$\delta = \frac{\sigma p d}{\nu \kappa T} + \frac{\sigma d^2 a^2}{3\kappa T} \cdot \frac{1}{(m/\mu + \Sigma 1/N_n)} \tag{17}$$

from which we may write for the reduced creep compliance

$$\bar{J} = (\text{creep compliance}) \cdot \nu \kappa T = p + \frac{1}{3(m/\mu + \Sigma {}^1/N_n)} \tag{18}$$

As before, no general explicit relations may be written, but again it is desirable to draw some general conclusions.

The value of the creep compliance will rise from its initial value

$$\frac{1}{\nu \kappa T} \left[p + \frac{a^2 d\nu}{3\,(m/\mu + \Sigma 1/(N_n)_0)} \right] \tag{19}$$

which, for this system is the elastic response to an absolute maximum of

$$\frac{1}{\nu \kappa T} \left[p + \frac{a^2 d\nu\mu}{3m} \right] \tag{20}$$

when the chains will have relaxed completely. In a model system, however, this maximum will not be achieved because, some time prior to this, the load will have distributed itself uniformly over all the chains which were initially shorter than μ monomer units, at which point creep will cease. Naturally, if $\mu = \infty$ or $m = 0$, creep will continue indefinitely.

The equations derived for creep are even more complicated to handle than those of relaxation; since they involve the solution of a system of linked, iterating differential equations. However, for a given (and small) number of saltating chains, a solution can be found, although it is still extremely unwieldy, and it appears at this juncture that no practical benefit will result from attempting to obtain approximate solutions.

Another method of handing this would be by inversion of the Prony series, remembering that the product of the Laplace transforms of the modulus and the creep functions equals 1[7].

7 It should be noted that the product of the relaxation modulus (Eq. 12) and the creep compliance (Eq. 18) is apparently 1. This, however, is not really the case, since $\Sigma(1/N_i)$ is different in the two cases because they are loaded differently

18. The Stress-Strain Curve

Writing $\sigma = M\epsilon$ and $M = f(Y\epsilon^2/R)$, we may differentiate to obtain

$$\frac{d\sigma}{d\epsilon} = M + 2\left(\frac{Y\epsilon^2}{R}\right) f'\left(\frac{Y\epsilon^2}{R}\right) \tag{21}$$

where the derivative f' is taken with respect to ϵ.

The general case of this expression cannot be given in closed form, but it is nevertheless possible to obtain the value of $d\sigma/d\epsilon$ for the cases for which M has been calculated as a function of $\epsilon Y \vartheta (= \epsilon^2 Y/R)$ by graphical methods, and then plot ϑ as a function of ϵ. This has been done in Fig. 21 for the model with $j = 6$. The curves obtained show, as expected from the basic assumptions for the model, a decreasing sensitivity to rate as the rate of extension increases. At very low rates, the curve appears to become almost parallel to the strain-axis; this is only an illusion because the gradient of the stress-strain curve at such slow rates is determined by the relative values of m and j, and the lengths μ and N_i. The particular case selected has a creep rate only slightly smaller than the strain rate at this level of stress.

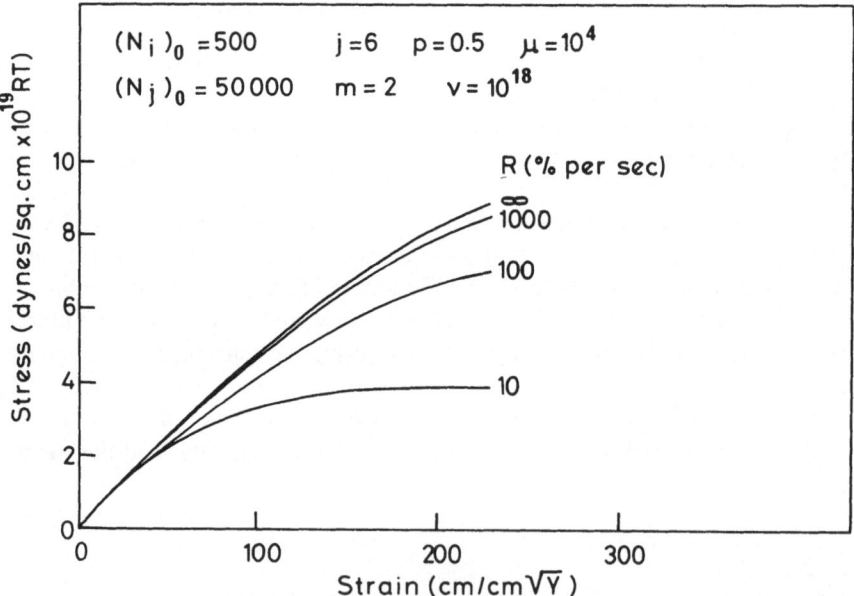

Fig. 21. Calculated stress strain curves at various rates of extension for model vulcanizate[77]

19. Saltation and Reinforcement

The way in which redistribution of load over individual chains occurs was described earlier for two individual chains. In the model under consideration, this can be best described by Fig. 22, which shows the initial distribution of load for the model $j = 48, m = 2, p = 0.5$ and $\mu = 10^4$, and the distribution as a function of the time-strain variable $Y \vartheta \epsilon$. If all chains are assumed to be equally strong, and if catastrophic failure is further assumed to proceed from the failure of one chain, the model vulcanizate will become stronger under instantaneous loading after relaxation in accordance with Fig. 22. However, due to the increasing lengths of the mobile chains, the fixed chains bear an increasing proportion of the load until they become the principal elements in the system; this is the point of maximum strength, and is a discontinuity on the curve.

In other configurations, it is possible that the failure of one element will not lead immediately to the failure of all the other elements. This would be the case of an even much larger difference between the lengths of the two shortest chains. Such a distribution, although stronger, must of necessity show considerably increased stress softening.

Sakurai and his associates[91] were able to demonstrate that the tensile stress in a strip of reinforced rubber held at constant extension for 900 sec was related to its initial tensile stress by a linear relation covering all extensions and loadings of each of the blacks studied. Similar relations for non-black fillers could not be drawn, as each filler loading resulted in a different line; these, however, were parallel. Starting with a given stress, ISAF showed the most drastic fall in stress ("relaxation"), followed by GPF, activated silica, FT and "silicate". The fact that more rapid relaxation appears to be related to increased reinforcement supports the main features of the model.

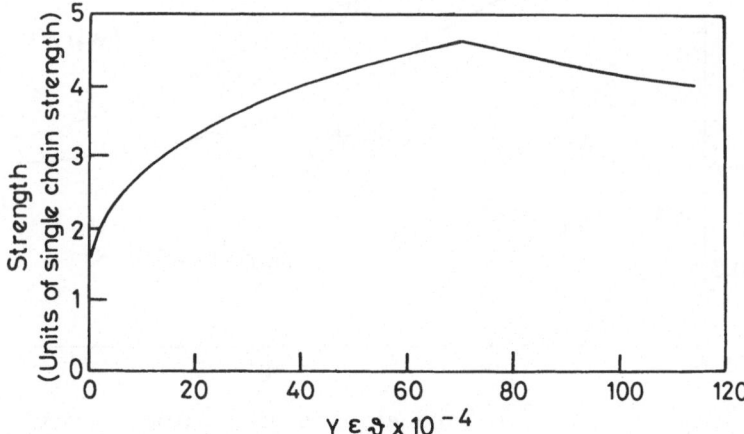

Fig. 22. Strength of model vulcanizate showing increase with time-strain variable until fixed chains assume maximum proportion of load

It is of particular interest to investigate the behaviour of a graphitised black in comparison to a standard black according to the model. According to the various sources noted graphitised black is characterized by: (1) lower viscosity in uncured mixes according to some authorities[10, 59] or higher viscosity according to others[8], possibly depending on the elastomer used; (2) somewhat reduced tensile strength of the vulcanizates[8, 59] and (3) much reduced abrasion resistance[10]; (4) lower exotherm on DTA[59]; (5) reduced-bound rubber content which does not however reach zero[85]; (6) reduced value of T_m on solvolysis[85]; (7) lower surface coverage of adsorbed species[71] but with the same order of chemisorbed molecules per basal plane; (8) lower modulus at any extension[18, 92] while the ratio of moduli of compound containing the standard black to those of compound with graphitized blacks increases with extension ratio[92]; (9) greater hysteresis[8]. In accordance with the concepts developed in this part of the review, such behaviour would imply lower adsorption energy and lower activation energy in the sense of Eq. (1). It should be noted that authors in Ref. 8 and Ref. 59 do not agree as to the relative losses of tensile strength and abrasion resistance on graphitization.

Equation (11) is convenient for a qualitative understanding of the situation. We now compare the behaviour of two macromolecules as discussed earlier (Sect. 14) in two similar cases, differing only in the value of E, which is included in the term. Note that because E appears as an exponent, it is the differences between two values of E which are of interest, not their ratios. For the purposes of calculation, a difference of $1.8 \kappa T$ has been adopted, and curves similar to W_1/W of Fig. 13 have been

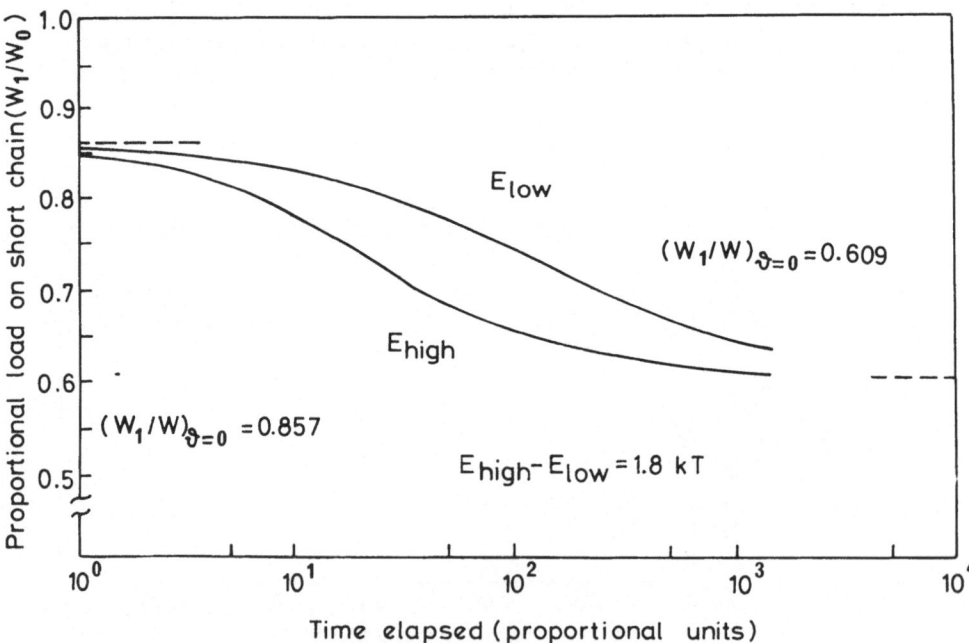

Fig. 23. Redistribution of load on to shorter of two chains ($N_{10} = 500$, $N_{20} = 3000$) for which the activation energy of saltation differ by $1.8 \, kT$ (logarithmic time scale)

drawn for the redistribution of load between the two chains in the case of high and low energy activation (Fig. 23). It will be seen that at any given time after the establishment of the extension, the load on the shorter chain had dropped *faster* in the case of the high activation energy, which results in greater reinforcement according to the concepts developed in this paper.

In considering abrasion, it will be noted that while actual abrasion loss is a phenomenon involving extreme properties at the surface, layers of vulcanizate immediately contiguous are subjected to many cycles of increasing stress before they are abraded, during which chain-length adjustment develops continuously.

20. Saltation and Strength

The previous section has shown how chain extension results in a redistribution of load between the various chains. We shall now attempt to calculate the influence of saltation on the tensile strength of the loaded vulcanizate, as first presented in Ref.[59].

The strength of each macromolecular backbone is characterized by the strength of the C–C or C=C bonds modified by the strains imposed by the side groups of the various polymers. The tension in the chain will reach this value when the extension of the stress-bearing section reaches a given fraction of its fully extended length. The stress-bearing section is that portion of a macromolecule between two crosslinks, a crosslink and an attachment to a carbon black aggregate or two such attachments. In the situation prevailing in the model previously described, the stress-bearing length of any given chain is N_i, and this will fail when it is extended so that the distance in monomer units measured in a straight line between its two anchor points becomes N_i/λ, $\lambda > 1$. At present we do not define λ.

The problem is now treated in the following manner: The model vulcanizate is subjected to a sudden extension, allowed to relax for a given period, and then subjected to additional steps of equal extension and equal periods of relaxation. In this way, an extension at a fixed rate is replaced by a series of steps, and the equivalent rate of extension will be $R = \epsilon/\vartheta$, where ϵ is the extension at each step and ϑ the corresponding relaxation period. It is seen that the number of steps taken is not a factor in this model, provided that ϑ is small.

The condition for rupture now requires [see Eq. (13)] that the i-th chain will fail when

$$\frac{a(N_i)_0}{\lambda} = \frac{(1-p)\,\epsilon d}{2} = \frac{1-p}{2}\,R\vartheta d \tag{22}$$

In our model of the black-polymer attachment, N_i also increases with time, leading to the following condition for rupture

$$(N_i)^2 = (N_i)_0^2 + \frac{ad\,Y\vartheta\epsilon}{D} = \lambda^2\,(1-p)^2\,\frac{d^2}{4}\cdot\frac{\epsilon^2}{a^2} \tag{23}$$

where D stands for the expression $p(m/\mu + \Sigma 1/N_i) + vda^2/3$. Equation (23) can now be rewritten as

$$\frac{1}{R} = \frac{D}{aY}\left[\frac{\lambda^2(1-p)^2}{4a^2} - \frac{(N_i)_0^2}{d^2\epsilon^2}\right] \tag{24}$$

At the rate R, the chain whose length is N_i monomer units will rupture when the complete unit cell reaches an elongation ϵ. The value of D is, of course, not a constant, but in any acceptable system, the distribution consists of at least a few short chains together with many longer ones, and D is not likely to vary rapidly when compared to the variations in N_i. D may therefore be considered constant for our present purposes. It is also assumed that failure occurs first in chains attached to a carbon black particle rather than a segment between crosslinks.

Some interesting observations can now be made, and these may be more easily visualized by reference to Fig. 24. The rate of extension at which the i-th chain will rupture is shown schematically as a function of the different variables involved. In Fig. 24a it is seen that if $\epsilon(1 - p)d/2$ is smaller than aN_i/λ, no rupture of the chain will take place no matter what rate of extension is applied. Other parameters remain-

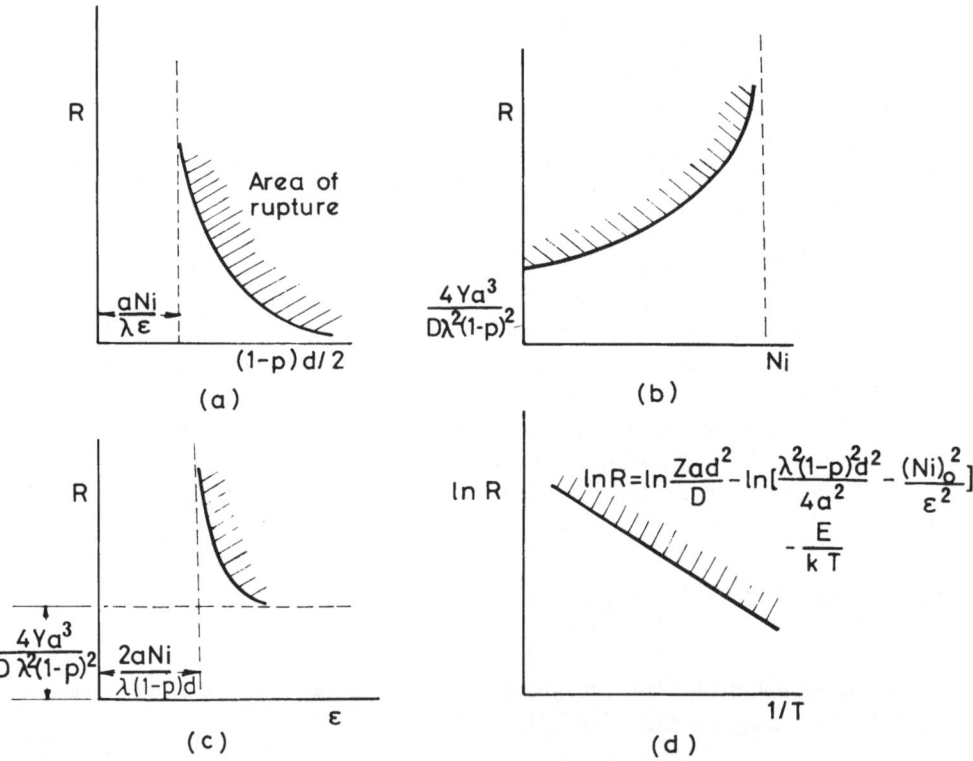

Fig. 24a–d. Maximum rates of strain possible with no chain rupture as a function of a cell geometry, b unstrained chain length, c maximum strain, d temperature [78]

ing constant, the rate at which rupture occurs falls continuously and rapidly as $\epsilon(1 - p)d/2$ rises above aN_i/λ, which is the same as saying that at any given rate, an increasing number of chains will fail as the elongation increases.

From 24b it would appear that even when $N_i = 0$ a finite rate of extension may be supported without rupture; this is of course a fallacy, since $D \longrightarrow \infty$ as $N_i \longrightarrow 0$. However, as N_i increases, the rate which will cause rupture increases rapidly, until a situation is reached at a value of $N_i = \epsilon\lambda(1 - p)d/2a$ at which no rupture will occur even at the greatest speeds. It is difficult to reconcile this conclusion with the instinctive picture which we have of the behaviour of vulcanizates. Nevertheless, omitting the case of a crosslink directly bonded to the carbon black surface, a case may be imagined in which the rate of increase of number of links, resulting from the forces generated by the applied extension, is greater than the rate of extension.

In 24c, we have a representation of the tendency to fail if extension is allowed to proceed to a fixed, final value. It will be seen that provided ϵ is below the value $2aN_i/\lambda d(1 - p)$, extension can take place at any rate, with no danger of rupture. However, as extension is taken further, the rate, above which rupture will take place at any given elongation, falls until it reaches a value determined by the other parameters.

The relationship between the maximum rate with no rupture, and the temperature of the model vulcanizate is obtained through $Y = Z \exp(-E/\kappa T)$, and in this case a simple shift factor applies, since for a given chain to rupture, all other parameters being equal,

$$ln \frac{R_1}{R_2} = \frac{E}{\kappa} \left(\frac{1}{T_2} - \frac{1}{T_1} \right) \tag{25}$$

This is shown graphically in Fig. 24d.

21. Discussion

The experiments described in the second half of this review were designed to support the contention that relative movement called saltation takes place between elastomer molecules and black carbon during stressing of a vulcanized, loaded elastomer. This saltation would relieve local high stresses and result in a distribution of stress which would effectively reinforce the vulcanizate.

If an analysis is made of many of the papers and surveys on the subject, it will be noted that there is actually very little reference made to reinforcement, as defined in this review. Very detailed consideration is given to the influence of structure on modulus, to aggregation and agglomeration, to voids and absorption and to occluded rubber. Polymer/surface interactions are treated, and there is discussion of reduced variable plots correlating strength and elongation in failure envelopes. Much effort has gone into studying viscoelastic behaviour and the influence upon it of certain carbon black parameters. However, all these have very little to do with the enhancement of tensile or tear strength or abrasion resistance. Kraus's excellent surveys[50, 51], for example, deal with strength only briefly. Harwood and Payne[40] do develop a failure criterion, but as has been shown above, this is based upon calculations which throw doubt upon its meaning. There is, in any case, little enough in their criterion to throw light on the mechanism of reinforcement, since apparently, according to these authors, all materials behave in much the same way.

Apparently, there is some misunderstanding about what happens to rubber during abrasion. The following statement by Donnet, Papirer and Vidal[23] is typical of many similar opinions: "Yet, in practice, filled rubbers such as used in tyres are, by far, not subjected to stresses reaching the high values required for rupture". This may be true for rubber compounds used for side wall construction, between plies and cushion. But it is definitely not the case for the most critical part of the tyre, the tread. Shallamach[83] has shown how the abrasion process involves the plucking of fibrils from the tread surface, and Ecker[24] has shown that a reciprocal relationship exists between abrasion loss and high speed tensile strength at 100 °C. It is the present author's contention that, in the last analysis, reinforcement is the result of the effect of carbon black on the extreme properties of the vulcanizate influenced of course by effects at lower extensions. As a consequence, all the studies made on rubber compounds at low strains, and on carbon black in relation to low strains, while being of tremendous importance for the provision of basic information, address themselves only in minor part to answering the question: how does carbon black act to reinforce rubber?

The author expresses a personal opinion that the slipping mechanism, first conceived by Houwink[44], taken up by Dannenberg[19] and given a mathematical treatment as a saltation process by Rigbi[77], is the only theory actually dealing with the problem.

Blanchard[4] and Blanchard and Parkinson[6] and later Mullins and Tobin[64] suggested that reinforcement could be explained by a mechanism involving the gradual softening of hard regions surrounding carbon black particles. This is a formalism for a model which does little to explain the mechanics of reinforcement.

Hess and coworkers[42] have demonstrated by means of excellent microphotography that separation between elastomer and carbon black occurs at high overall strains which, of course, result in particularly high stresses at the extremities of the black particles. While claiming that the separation is not an artifact, Hess admits[41] that the frequency of the resulting vacuoles may have been increased by the presence of the rigid supports required for the technique. This observed separation obviously lessens the benefits from saltation, and the curves of Fig. 24 will be changed qualitatively accordingly, but separation does not explain reinforcement. It merely indicates that the more reinforcing the black, the less separation is to be observed.

Dannenberg[19] summarized the phenomena which must be explained by a theory of reinforcement. One of these is the reversed modulus-temperature dependence shown by the rubber on the addition of carbon black, and is closely linked to the observations by Oono et al.[66]. The modulus of crosslinked, unfilled rubber increases with temperature; addition of carbon black reduces this tendency until, at sufficient concentration, the modulus-temperature gradient is reversed. This effect may be explained qualitatively by the saltation mechanism: the more rapid thermal motions responsible for increased modulus in the rubber chains are also responsible for greater tension and therefore for greater saltation leading to longer chains, the rate of saltation under stress increasing with temperature as well. While this description appears reasonable, it still awaits mathematical expression.

22. Desiderata for an Ideal Reinforcing Filler: Speculations

The above considerations may now enable us to define certain features desirable in a filler which would give maximum reinforcement. Given an elastomer, crosslinked to a given degree, a filler will be reinforcing if it allows the elastomer to increase in strength. This strength is derived from the ability of the filler to redistribute the stresses along the macromolecules till they become uniform, when the complete vulcanizate would achieve its greatest strength in any mode of loading. The most highly reinforcing fillers are those which will show greatest stress softening due to saltation and least softening due to rupture, and this condition will result from minimum surface energies provided that they are sufficient for the black to maintain contact with the polymer molecules. However, from a technological point of view, it is often necessary to provide vulcanizates exibiting a high modulus, and as this necessitates a large number of attachments. Therefore compromise desiderata must be proposed as follows:

The *smaller* the activation energy, the greater the reinforcing ability of a black, provided that it has a sufficiently increased number of absorption sites to maintain a constant modulus[8].

Acknowledgements. The author is grateful to a number of persons and organizations for having made facilities available or giving him time at various dates over the last twenty three years to pursue his personal interest in carbon black which has resulted in this paper. In particular, he wishes to thank the Rubber Research Association Ltd. of Haifa; Technion-Israel Institute of Technology, Haifa; Scientific Department, Ministry of Defence, Tel-Aviv; Cabot Corporation, Boston, Mass.; CNRS- Centre de Recherches sur la Physico-Chimie des Surfaces Solides, Mulhouse, as well as many friends, colleagues and correspondents too numerous to name.

---

8   Following the preparation of this paper, the author's attention was drawn to a paper presented by D. C. Edwards at the 1978 A.C.S. Rubber Division meeting in Boston. The evidence presented supports the mechanism developed here, and indicates that his work was planned and successfully concluded with these speculations in mind. Later discussions with S. Wolff make it appear likely that the function of the silane (produced commercially as Si69) in increasing the reinforcing activity of silica in rubber is to provide a mechanism similar to the one described

23. References

1. Ban, L. L., Hess, W. M.: Current progress in the study of carbon black microstructure and general morphology. In: Renforcement des elastomères (J.-B. Donnet, ed.) p. 81, Paris: CNRS 1975
2. Ban, L. L., Hess, W. M., Papazian, L. A.: Rubber Chem. Technol. *47*, 858 (1974)
3. Beebe, R. A. et al., quoted by Holmes, J. M. in (Flood, E. A., ed.): The solid-gas interface, Ch.5. New York: Dekker 1967. In particular, see Fig. 5–2 on p. 132
4. Blanchard, A. F.: J. Polym. Sci. Al, *8*, 813, 835 (1970)
5. Blanchard, A. F.: The immobilization of elastomers on filler surfaces. In: Renforcement des elastomères (J. B. Donnet, ed.) p. 41, Paris: CNRS 1975
6. Blanchard, A. F., Parkinson, D.: Ind. Eng. Chem. *44*, 799 (1952)
7. Boonstra, B. B., Taylor, G. L.: Rubber Chem. Technol. *38*, 943 (1965)
8. Boonstra, B. B.: J. Appl. Polym. Sci. *11*, 389 (1967)
9. Boonstra, B. B., Medalia, A.: Rubber Age *92*, 892 (1962)
10. Brennan, J. J., Jermyn, T. E., Boonstra, B. B.: J. Appl. Polym. Sci. *8*, 2687 (1964)
11. Brennan, J. J., Dannenberg, E. M., Rigbi, Z.: Some aspects of stress softening in reinforcement by carbon black. Proc. Fifth Internat. Rubber Conf. Brighton/England, 1967, p. 123, London: Mclaren 1968
12. Brennan, J. J., Jermyn, T. E., Perdigao, M. F.: Influence of carbon black on stress softening (Mullins effect). Paper no. 36, Division of Rubber Chemistry, ACS, Detroit, Michigan, 1964
13. Buchan, S., Rae, W. D.: I.R.I. Trans. *21*, 323 (1946)
14. Bueche, F.: J. Appl. Polym. Sci. *7*, 1165 (1963)
15. Bueche, A. M., White, A. V.: J. Appl. Phys. *27*, 980 (1956)
16. Chasset, R.: Relaxation viscoelastique de vulcanizats d'un copolymere butadiene-styrolene (SBR-1500) contenant differentes proportions d'un noir de carbone ISAF. In: Renforcement des elastomères. (J. B. Donnet, ed.) p. 193, Paris: CNRS 1975
17. Cotten, G. R., Boonstra, B. B.: J. Appl. Polym. Sci. *9*, 3395 (1965)
18. Cotten, G. R. et al.: Kautsch. Gummi-Kunst. *22*, 477 (1969)
19. Dannenberg, E. M.: I.R.I. Trans. *22*, T26 (1966)
20. Dannenberg, E. M., Brennan, J. J.: Rubber Chem. Technol. *39*, 597 (1966)
21. Dijauw, L. K., Gent, A.: J. Polym. Sci. (Symposia) *48*, 159 (1974)
22. Donnet, J. B., Voet, A.: Carbon black – physics, chemistry and elastomer reinforcement. New York: Dekker 1976
23. Donnet, J. B., Papirer, E., Vidal, A.: Nippon Gomu Kyokaishi *50*, 223 (1977)
24. Ecker, R.: Rubber Chem. Technol. *39*, 823 (1966)
25. Eirich, F. R.: Rubber reinforcement in perspective. In: Renforcement des elastomères. (J. B. Donnet, ed.) p. 1, Paris: CNRS 1975
26. Eyring, H., Eyring, E. M.: Modern chemical kinetics. Chapter on transport mechanics, New York: Reinhold 1963
27. Flory, J. P., Rehner, J.: J. Chem. Phys. *11*, 512 (1943)
28. Gent, A. N.: J. Appl. Polym. Sci. *18*, 1397 (1974)
29. Gessler, A. M.: Rubber Age *101*, 54 (1969)
30. Glucklich, J., Landel, R. F.: J. Appl. Polym. Sci. *20*, 121 (1976)
31. Greensmith, H. W., Mullins, L., Thomas, A. G.: The strength of rubbers. In: Chemistry and physics of rubberlike substances (Bateman, L., ed.), p. 249, London: Maclaren 1963
32. Gregory, M. J., Metherell, C., Smith, J. F.: Plastics and Rubbers: Mater. Appl. *1978*, 37
33. Gurney, W. A.: Trans. I.R.I. *21*, 31 (1945)
34. Guth, E., Gold, O.: On the hydrodynamical theory of viscosity of suspensions (Abstract), Phys. Rev. *53*, 322 (1953)
35. Halpin, J. C.: J. Appl. Phys. *35*, 3133 (1964)
36. Halpin, J. C.: Rubber Chem. Technol. *38*, 1007 (1965)
37. Halpin, J. C., Bueche, F.: J. Appl. Phys., *35*, 3142 (1964)
38. Harwood, J. A. C.: J. Appl. Chem. *17*, 533 (1967)

39. Harwood, J. A. C., Mullins, L., Payne, A. R.: J. Appl. Polym. Sci. *9*, 3011 (1965)
40. Harwood, J. A. C., Payne, A. R.: J. Appl. Polym. Sci. *10*, 315 (1966)
41. Hess, W. M.: Private communication
42. Hess, W. M., Marsh, P. A.: Norelco Reporter *12*, 1 (1965)
43. Hobden, J. F., Jellinek, H. H.: J. Polym. Sci. *11*, 365 (1953)
44. Houwink, R.: Rubber Chem. Technol. *28*, 888 (1958)
45. Kamenskii, A. N. et al.: Mekh. Polimerov *3*, 291 (1967), translated as: Electron microscope investigation of the type of fracture of filled rubbers. Polymer Mech. *2*, 198 (1967)
46. Kaufman, S., Slichter, W. P., Davis, D. D.: J. Polym. Sci., (*A2*)*9*, 829 (1971)
47. Knauss, W. G.: The time-dependent fracture of viscoelastic materials. In: Proc. Internat. Conf. Fracture, Sendai, Japan, 1965, 1139
48. Kolthof, I. M., Gutmacher, R. E., Kahn, A.: J. Phys. Chem., *55*, 1240 (1951)
49. Kraus, G. (ed.): Reinforcement of elastomers, New York, Wiley, 1965
50. Kraus, G.: Adv. Polym. Sci. *8*, 155 (1971)
51. Kraus, G.: Angew. Makromol. Chem. *60/61*, 215 (1977)
52. Kraus, G., Dugone, L.: Ind. Eng. Chem. *47*, 1809 (1955)
53. Kraus, G., Childers, C. W., Rollman, K. W.: J. Appl. Polym. Sci. *18*, 229 (1966)
54. Kraus, G., Gruver, J. T.: Rubber Chem. Technol. *41*, 1256 (1968)
55. Kuhn, W.: Kolloid-Z. *76*, 258 (1936)
56. Lake, G. J., Lindley, P. B., Thomas, A. G.: Fracture mechanics of rubber. In: Fracture 1969 (Pratt, P. L., ed.), London: Chapman and Hall 1969
57. Lambert, D. H.: Internal Report, Cabot Corp., Billerica, Mass.
58. Le Bras, J., Papirer, E.: J. Appl. Polym. Sci. *22*, 525 (1978)
59. Lezhnev, N. N. et al.: Dokl. Akad. Nauk. SSSR. *160*, 861 (1965), translated as: Investigation of the properties of the structures of rubbers strengthened by interaction with carbon black, Doklady Phys. Chem. *161*, 107 (1965)
60. Mason, P.: J. Appl. Phys. *29*, 1146 (1958)
61. Medalia, A. I.: Rubber Chem. Technol., *45*, 1171 (1972)
62. Medalia, A. I.: J. Colloid Interface Sci. *32*, 115 (1970)
63. Medalia, A. I.: Rubber Chem. Technol. *51*, 437 (1978)
64. Mullins, L., Tobin, N.: J. Appl. Polym. Sci. *9*, 2993 (1965)
65. O' Brien, J. et al.: Macromol. *9*, 653 (1967)
66. Oono, R., Ikeda, H., Todani, Y.: Angew. Makromol. Chem. *46*, 47 (1971)
67. Papirer, E., Nguyen, V. T., Donnet, J.-B.: J. Polym. Sci. (Polym. Letters), in press
68. Payne, A. R., Whittaker, R. E.: Importance of hysteresis in the reinforcement of elastomers. In: Renforcement des elastomères (Donnet, J.-B., ed.), p. 233, Paris: CNRS 1975
69. Peremsky, R.: Kauc. Plast. Hmoty *1963*, 37
70. Pliskin, I., Tokita, N.: J. Appl. Polym. Sci. *16*, 473 (1972)
71. Rivin, D.: Rubber Chem. Technol. *44*, 307 (1971)
72. Rivin, D., Aron, J., Medalia, A. I.: Rubber Chem. Technol. *41*, 330 (1968)
73. Rivlin, R. S., Thomas, A. G.: J. Polym. Sci. *10*, 291 (1953)
74. Rigbi, Z.: J. Appl. Polym. Sci. *12*, 2736 (1968)
75. Rigbi, Z.: Kolloid Z. u. Z. Polymere *224*, 46 (1968)
76. Rigbi, Z.: ibid. *225*, 40 (1968)
77. Rigbi, Z.: ibid. *223*, 127 (1968)
78. Rigbi, Z.: Rev. Gen. Caoutch. Plast. *45*, 625 (1968)
79. Rigbi, Z.: Bull. Res. Counc. Israel *6C*, 67 (1957)
80. Rehner, J.: The nature of polymer-filler attachmets. In: Reinforcement of elastomers (Kraus, G., ed.) New York: Wiley Interscience 1965
81. Sambrook, R. W.: J. Inst. Rubber. Ind. *1970*, 210
82. Schoon, Th. G. F., Adler, K.: Kautsch. Gummi – Kunst. *19*, 414 (1966)
83. Shallamach, A.: Wear *1*, 384 (1958)
84. Silberberg, A.: J. Phys. Chem. *66*, 1872 (1962)
85. Sircar, A. K., Voet, A.: Rubber Chem. Technol. *43*, 973 (1970)
86. Smit, P. P. A.: Rheol. Acta *5*, 277 (1966)

 87. Smith, T. L.: J. Polym. Sci. *A 1*, 3597 (1963)
 88. Smith, T. L.: J. Appl. Phys. *35*, 27 (1964)
 89. Smith, T. L.: Strength and extensibility of elastomers. In: Rheology, Vol. 5 (Eirich, F. R., ed.), pp. 127–221, New York: Academic Press 1969
 90. Studebaker, M. L.: Rubber Chem. Technol. *30*, 1401 (1957)
 91. Sukurai, M. et aL: Repts. Progr. Polym. Phys. Japan *8*, 237 (1965)
 92. Taylor, G. L.: Internal Report Cabot Corporation *1963*
 93. Treloar, L. R. G.: The Physics of rubber elasticity, 2nd ed. Oxford: Oxford Univ. Press 1958
 94. Ulmer, J. D., Hess, W. M., Chirico, V. E.: Rubber Chem. Technol. *47*, 729 (1974)
 95. Takayanagi, M., Minami, S., Uemura, S.: J. Polym. Sci. (Symposia), *C5*, 113 (1964)
 96. van Ooij, W. J.: Rubber Chem. Technol. *51*, 52 (1978)
 97. Wake, W. C.: Surface mobility and adhesion. Renforcement des Elastomères (Donnet, J.-B., ed.), p. 277, Paris: CNRS 1975
 98. Wall, F. T.: J. Chem. Phys. *11*, 512 (1943)
 99. Westlinning, H.: Kautsch. Gummi-Kunst. *20*, 5 (1967)
100. Williams, M. L., Landel, R. F., Ferry, J. D.: J. Amer. Chem. Soc. *77*, 3701 (1955)

Received November 14, 1979
H.-J. Cantow (editor)

Transformations of Phenolic Antioxidants

and the Role of Their Products in the Long-Term Properties of Polyolefins

Jan Pospíšil

Institute of Macromolecular Chemistry, Czechoslovak Academy of Sciences,
162 06 Prague 6, Czechoslovakia

Transformation mechanisms of the most important classes of phenolic antioxidants are given for the conditions of inhibited oxidation. Properties and importance of the main types of formed products are reviewed with respect to the long-term properties of polyolefins.

Table of Contents

I. Introduction

Utility properties of polyolefins irreversibly degenerate under environmental effects. The thermal oxidation processes take part above all during processing while the photoinduced processes contribute to the entire degradation mainly in atmospheric ageing. The perfect knowledge of individual degradation mechanisms and of their specific role under given conditions and the optimum utilization of all aspects of the mechanism of the stabilizers' action are the conditions for the effective stabilization of polyolefins. To describe the main processes undergon by a polymer (RH) in the atmospheric ageing, Scheme 1 may be employed according to[1]. The radicals R' arising by the thermal or photoinduced initiation participate in a chain autooxidation reaction[2-4] in the

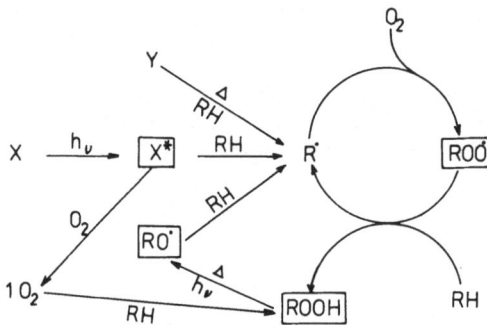

Scheme 1

propagation steps:

$$R' + O_2 \longrightarrow RO'_2 \tag{1}$$
$$RO'_2 + RH \longrightarrow ROOH + R' \tag{2}$$

The resulting hydroperoxide is decomposed in a monomolecular (3a) or bimolecular (3b) reaction

$$ROOH \longrightarrow RO' + HO' \tag{3a}$$
$$2\,ROOH \longrightarrow RO' + RO'_2 + H_2O \tag{3b}$$

The radicals RO'_2 and RO' are deactivated by chain-breaking antioxidants[5] (these are marked InH and the derived phenoxyl radicals In' throughout the text)

$$RO'_2\ (RO') + InH \longrightarrow ROOH\ (ROH) + In' \tag{4}$$

Preventive antioxidants[3], comprising light stabilizers, hydroperoxide decomposers, and deactivators of metals, take part in the deactivation of further species shown in Scheme 1.

This review analyzes the data on most important antioxidants for polyolefins, i. e. the data on phenols. To learn consistently the whole mechanism of polyolefin stabilization, it is not possible to consider only the facts about the kinetics of the process and relationships between the chemical structure of stabilizers and their observed efficiency. It is necessary to understand the mechanism of stabilizer action on the basis of knowledge of transformations which occur in the inhibited oxidation and of properties of resulting products. The product analyses were obtained above all from the study of models and from independent syntheses. The confirmation of results under real conditions cannot be always carried out consistently because of low concentrations, difficult isolation, and reactivity of transformation products. It is also difficult to analyze such reaction mixtures directly in a polymer[6].

Earlier data on transformations of phenolic antioxidante have been compiled in reviews[3, 4, 7]. This article deals with the transformations of typical and technically important phenolic antioxidants and includes the literature data available until end of April, 1979. The emphasis is laid on transformations occurring under the conditions which simulate an inhibited thermal oxidation and photochemical processes during ageing of polyolefins. The chemical and photochemical properties of main products and available data concerning the subsequent transformations and the effects of transformation products on the oxidation of hydrocarbon systems are included.

II. Monohydric Phenols

Mononuclear and multinuclear monohydric phenols are the technically most applied antioxidants for polyolefins. Most of them are indicated as nonstaining. They are the typical representatives of chain-breaking antioxidants[5], acting according to (4) and fundamental relationships between their structure and efficiency were defined[3, 4, 8, 9]. They give rise to the phenoxy radical In' during the stabilization process and all further transformations occur via phenoxyl.

A. Phenoxy Radicals and Their Phenolic Coupling Products

The chemistry of phenoxyls is reviewed in[10–13]. They are scavengers of aliphatic radicals $R^{\cdot}$[14]. Although the presence of $R^{\cdot}$ in an oxidized polymer is of little probability at a sufficient pressure of oxygen, this reaction cannot be entirely neglected in a complex real system[15].

Phenoxyls undergo chemical transformations and react either as O-radicals I or as C-radicals (i. e., cyclohexadienonyls, II).

The stability and reactivity of phenoxyls is determined above all by steric effects of substituents R^1 to R^3 [12, 16] and by the extent of delocalization of the unpaired electron. Phenoxyls are able to cause the one-electron transfer[17, 18] and, consequently, the homolysis of O–H, N–H, or S–H bonds and even of aliphatic bonds C–H in extreme cases. Concerning the stabilization of polymers, these processes are

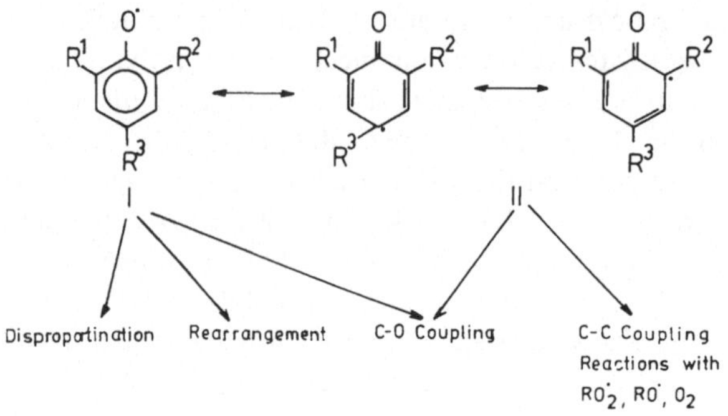

O·   O   O

R¹ ⌬ R²   ↔   R¹ ⌬ R²   ↔   R¹ ⌬ R²

R³   R³·   R³

I   II

Dispropartination   Rearrangement   C-O Coupling   C-C Coupling

Reactions with

RO·₂, RO·, O₂

Scheme 2

mostly undesired because they lead to the inefficient transformations of phenolic chain-breaking antioxidants or to the transfer of the autooxidation chain. In the first place, phenols which form reactive phenoxyls able to take part in reactions (5) and (6),

$$In^· + ROOH \rightarrow InH + RO^·_2 \qquad\qquad (5)$$

$$In^· + RH \rightarrow InH + R^· \qquad\qquad (6)$$

have to be excluded from application.

The risk of these reactions becomes evident particularly at high temperatures of oxidation[19, 20]. Phenoxyls derived from sterically hindered phenols have an optimum reactivity in terms of the stabilization process. It was shown by correlation of the antioxidation efficiency of various phenols with the relative reactivity of derived phenoxyls[21] that more stable phenoxyls are formed from more efficient antioxidants. The highly reactive phenoxyls may play a positive role only in homo-synergistic mixtures of phenols with sterically hindered phenols[5].

Phenoxyls formed from substituted 2,2' − or 4,4'-biphenyldiols are very stable. The unpaired electron is delocalized in these radicals over both aromatic nuclei[22, 23]. The resonance effect between aromatic nuclei of antioxidants having the structure of alkylidenebisphenols is interrupted by the carbon-atom bridge. The extent of unpaired electron delocalization is reduced in this way and the phenoxyls formed have characteristics similar to those of mononuclear phenols.

Under the conditions of polymer stabilization the typical reactions of phenoxyls are those (Scheme 2) in which only I or II take part, i. e. coupling via C−O or C−C, formal rearrangement connected with formation of benzyl radical (Chap. II A), or disproportionation (see the Chap. II B), on the one hand, and the reactions with radicals RO·₂, RO·, or with oxygen (see the Chap. II C) on the other. The transformation of phenoxyl and the final composition of the transformation products are affected very markedly by conditions of inhibited oxidation.

Phenoxyls, having bulky substituents in the positions 2, 4, and 6 (e. g., tert-butyls), do not dimerize[24]. If some of the substituents is less sterically demanding (this may be also the case of bridge in multinuclear phenols), C−C or C−O couplings take place. The mechanism of dimerization of sterically unhindered phenoxyls was studied[25] and the absolute rate constants of this reaction were determined. The results are important from the standpoint of antioxidation-effective homosynergistic mixtures of sterically hindered and unhindered phenols. According to[26], more frequent is a C—O coupling and the formation of cyclohexadienone derivatives of type III. The C—O coupling is reversible[27] and the equilibrium is influenced by steric effects of all substituents. The C—O coupling occurs also with multinuclear antioxidants, e. g., in the oxidation of 2,2'-dihydroxy-1,1'-binaphthyl[28].

III   IV

V

Cyclohexadienones of the type IV (n = 0, 1, 2)[29] are formed from 2,2'-biphenyldiols, 2,2'-methylenebisphenols or 2,2'-ethylenebisphenols via intramolecular C—O coupling. 5, 7, 3', 5'-Tetra-tert-butylspiro [2,3-dihydrobenzofuran-2,1'-cyclohexa-3', 5'-diene-2'-one] (IV, $R^1 = R^2$ = tert-Bu, n = 1) is stable. The similar compound IV (R^1 = tert-Bu, R^2 = Me, n = 1), prepared from the technically important antioxidant 2,2'-methylenebis (4-methyl-6-tert-butylphenol) by oxidation with sodium hypochlorite[30] is oily and crystallizes in the form of Diels-Alder addition compound V.

The dimers of type IV do not affect the course of tetrahydronaphthalene oxidation[31, 32]. They may be considered as inert admixtures from the view of thermal oxidation of polymers.

A more complex course of C—O coupling was observed with 2,4-di-tert-butylphenol[33], where the dimer VII is rapidly isomerized to the phenol derivative VIII which undergoes further oxidation.

Intermolecular C—C coupling occurs above all in the oxidation of antioxidants with a structure of 2,4— or 2,6-dialkylphenols[33—37]. 4, 6, 4', 6'-Tetra-tert-Butyl-2,2'-biphenyldiol X is formed from 2,4-di-tert-butylphenol[33] (Scheme 3). Cyclohexadienone XVIII is formed from 2,6-di-tert-butylphenol via XVII and rapidly isomerized to 4,4'-biphenyldiol XIX, which is further oxidized to diphenoquinone XX[34—36, 38] (Scheme 4). 2,6-Xylenol is similarly changed during retardation of auto-oxidation of benzaldehyde to tetramethyldiphenoquinone[37]. Cyclohexa-

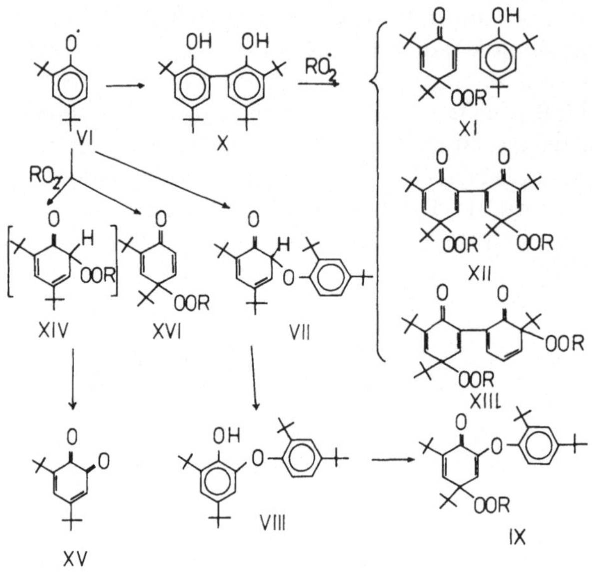

Scheme 3

dienones of type **XVIII** may be formed by the C—C coupling also from 2,4,6-trisubstituted phenols which have a non-bulky and readily cleavable substituent in the position 4[12].

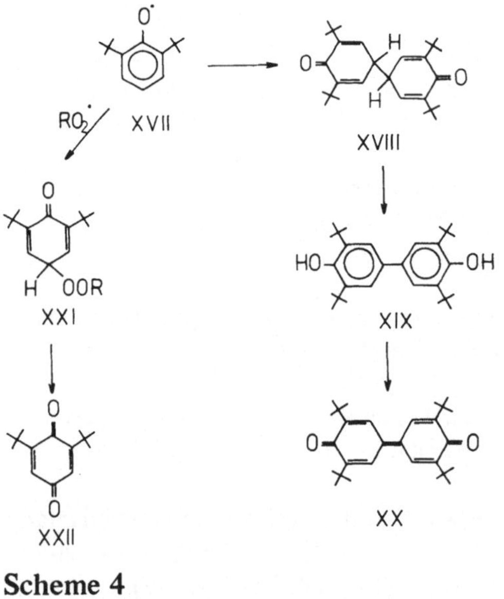

Scheme 4

The C—C coupling proceeds intramolecularly in 4,4'-alkylidene-bisphenols, which contain the bridge with a quaternary carbon atom[39, 40], under formation of cyclohexadienone **XXIII**.

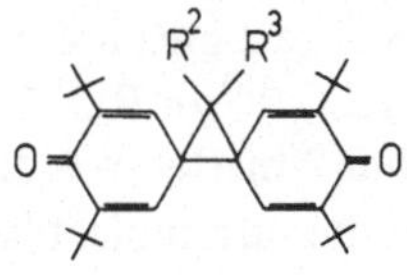

XXIII

The parallel C–C and C–O couplings take place via phenoxyl in the photo-oxidation of 2,4-disubstituted phenols: 4,4'-dimethoxy-6,6'-di-tert-butyl-2,2'-biphenyldiol and 2-tert-butyl-4-methoxy-6-(2-tert-butyl-4-methoxyphenoxy)phenol were formed from 2-tert-butyl-4-methoxyphenol[41].

Also the formally radical rearrangement of phenoxyl to a benzyl radical and the subsequent dimerization of the latter are often considered to be an alternative path in addition to the disproportionation to quinone methide[34, 42 –45]. However, the benzyl radical has not yet been with certaints proved[46, 47]. For example its origin allows the formation of 4,4'-ethylenebis(2,6-di-tert-butylphenol) XXVIII from

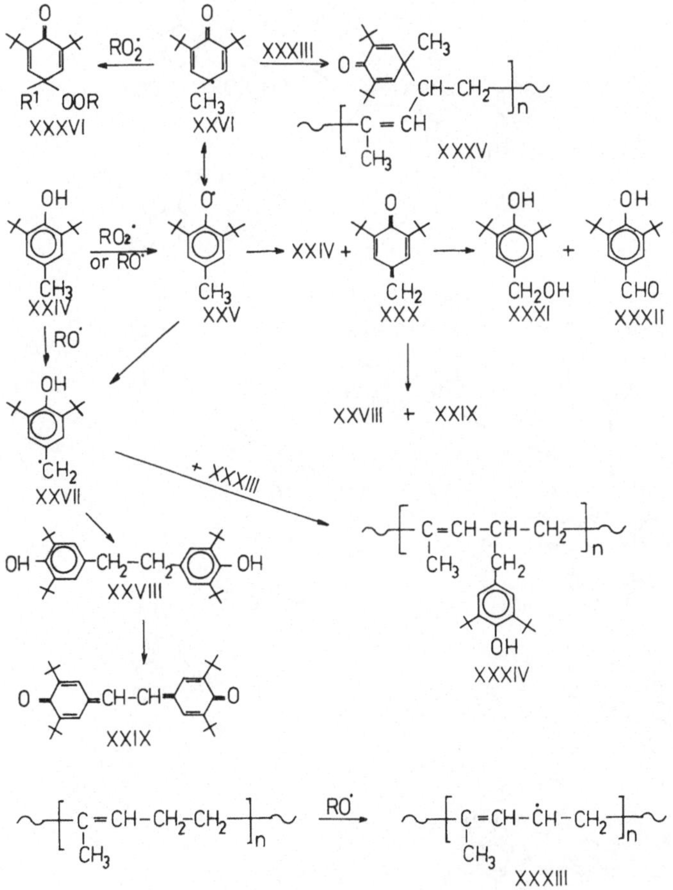

Scheme 5

2,6-di-tert-butyl- 4-methylphenol (BHT, **XXIV**) to be demonstrated during the action
of $RO_2^{\cdot}$[48] or $RO^{\cdot}$ [44] (Scheme 5) or of 2,2'-ethylene-bis (4,6-di-tert-butylphenol)
from 2-methyl-4,6-di-tert-butylphenol[29] (the reaction proceeds in this case further
via the intramolecular C–O coupling with spirocyclohexadienone **IV** ($R^1 = R^2$ = tert-
Bu, n = 2), which is the final product).

The transient formation of benzyl radical **XXXIX** has been assumed even in
oxidation of 2,2'-methylenebis (4-methyl-6-tert-butylphenol) **XXXVII** by radicals tert-BuO˙ or tert-BuO$_2^{\cdot}$[49]. A phenolic dimer **XL** and a trimer **XLVI**
arise as the main reaction products with a molar ratio of biphenol **XXXVII** to an
oxidation agent of 1–2 : 1. The presence of further two phenolic compounds and
several dark-coloured products, which strongly discolour the reaction mixture, was
ascertained in the reaction mixture by means of thin-layer chromotography (TLC).
The Scheme 6 was proposed to describe the oxidation transformations of **XXXVII**
under conditions simulating the inhibition process. It comprises (a) the formation of
phenoxyl **XXXVIII**, its rearrangement to the benzyl radical **XXXIX** and dimerisation
to **XL**, (b) the direct formation of **XXXIX** followed by dimerization, and (c) the dis-

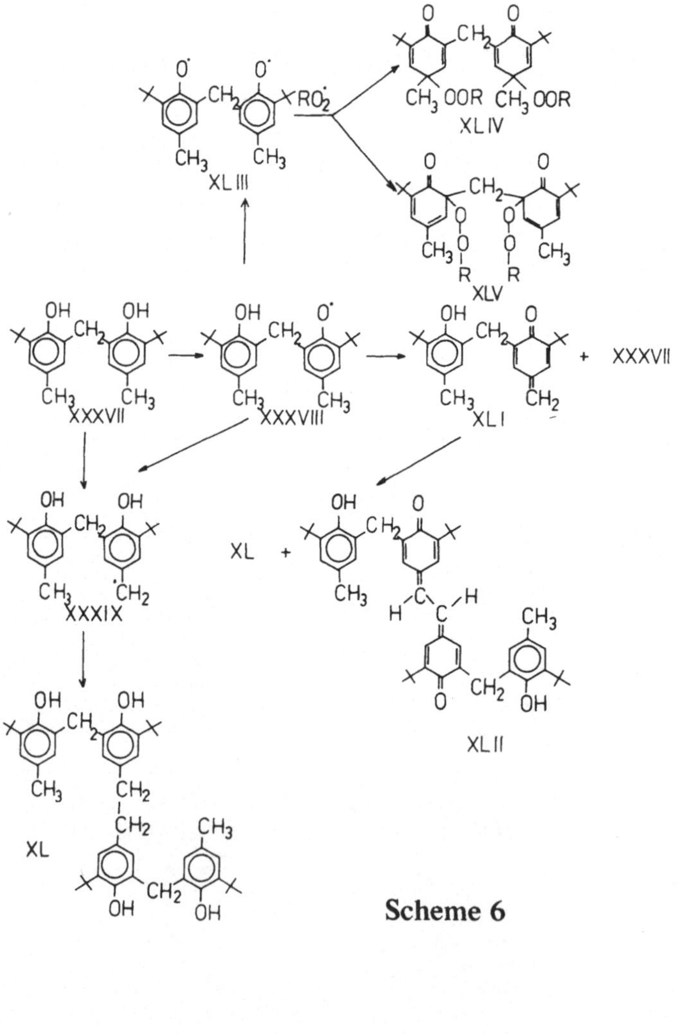

Scheme 6

proportionation of **XXXVIII** to quinone methide **XLI** and the initial bisphenol **XXXVII**; the transformation of **XLI** yields a mixture of **XL** and of a quinone methionide derivative **XLII**. The analogous sequence of reactions may explain the formation of the phenolic trimer **XLVI** which was isolated, or of higher oligomers, particularly tetramer, which presence in the reactio mixture is presumed on the basis of results of gel permeation chromatographic (GPC) analyses.

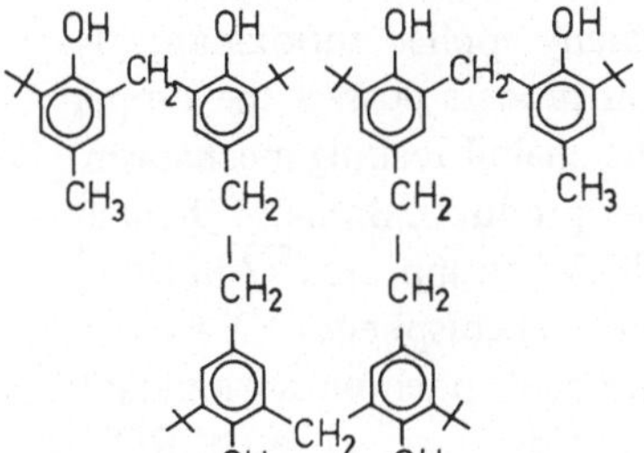

XLVI 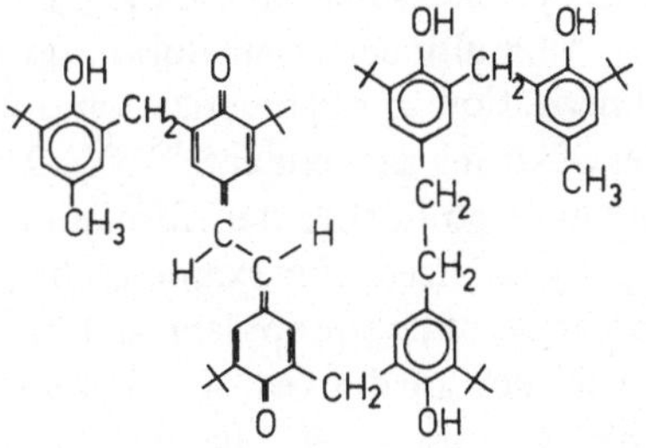   XLVII

The benzyl radical presumably participates also in an intermolecular oxidative coupling of another antioxidant, 4,4.-isopropylidenebis(2-methyl-6-tert-butyl-phenol). The resulting phenolic oligomer is further oxidized and forms the more complex product **XLVIII** via intramolecular C—O bonding[50].

Isolation of 2,6-di-tert-butyl-4-(3,5-di-tert-butyl-4-hydroxybenzyl)-4-methyl-2,5-cyclohexadienone **XLIX** from the oxidation of **XXIV**[51, 52] appear to prove the formation of the benzyl radical **XXVII**.

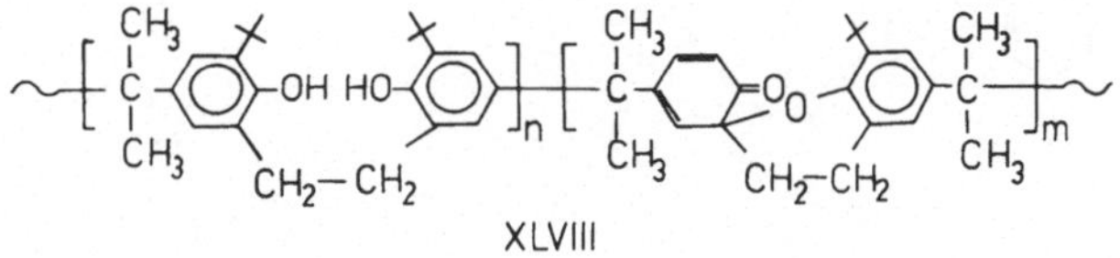

XLVIII

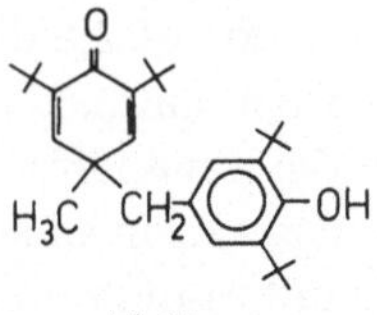

XLIX

The formation of benzyl radical **XXVII** from **XXIV** and its incorporation in to a macromolecule of natural rubber is claimed on experimental basis in[53]. About 20 % of used **XXIV** was bonded to the polymer chains in the optimum case forming a product of the assumed structure **XXXIV**. A considerable part of used BHT was transformed to ethylenebisphenol **XXVIII** and stilbenequinone **XXIX** during the reaction. Further products were 2,5-di-tert-butyl-4-hydroxybenzyl alcohol XXXI and the corresponding aldehyde XXXII, which were formed probably via quinone methide XXX. A model reaction proved that BHT is not bonded to polymer through the oxygen atom of phenoxyl. The mechanism formulated in[53] (Scheme 5) ascribes the origin of phenoxyl XXV as well as of alkyl radicals XXXIII leading to the formation of XXXIV to the action of RO˙ from the initiation system. However, in accordance with[54], neither abstraction of hydrogen from unsaturated polymer skeleton by phenoxyl XXV nor the consecutive reaction of polymer radical XXXIII with cyclohexadienonyl XXVI, giving the structure XXXV can be excluded.

All multinuclear phenols resulting from the originally applied antioxidants by dimerization or oligomerization during the inhibition process possess the properties of strong antioxidants[55-57]. They also act by the chain-breaking mechanism and undergo further transformations to more complex products. Some of them are rapidly oxidized. For example, the phenols XL and XLVI disappeared[55] in the oxidation of polypropylene at 180 °C faster than methylenebisphenol XXXVII, from which they are derived. At all events, the transformations of phenolic antioxidants, which give rise to the phenolic coupling products due to the deactivation of RO˙ or RO˙$_2$ radical are among the processes which positively affect the overall value of their stabilization capacity.

B. Quinone Methides

Sterically hindered phenoxyls having an alkyl on α-hydrogen atoms in the para position disproportionate readily to the initial phenol and quinone methide L[48, 58, 59]. This disproportionation is an irreversible process.

The disproportionation of 2,6-di-tert-butyl-4-methyl-phenoxyl XXV, which is derived from an important antioxidant 2,6-di-tert-butyl-4-methylphenol XXIV, has been discussed from the standpoint of kinetics and mechanism for several years[60-63]. XXV disappears in an aprotic medium by the 2nd order reaction[63] and a mixture of 2,6-di-tert-butylquinone methide XXX and phenol XXIV results (Scheme 5). The quinone methide XXX is very reactive and behaves according to[64] as a biradical. Coupling reactions occur and, consequently, ethylenebisphenol XXVIII and a dimeric quinone methionid compound 3,5,3', 5'-tetra-tert-butyl-4,4'-stilbene-quinone XXIX, are formed[65]. The last compound may result also in the reaction of XXX with hydroperoxides[66]. The details of origin of XXVIII are not entirely elucidated. Probably, the above mentioned 3,5-di-tert-butyl-4-hydroxy-benzyl XXVII takes part in the process[48, 67, 68] (see the Chap. II A.).

2,6-Diphenyl-4-methylphenoxyl disproportionates similarly to XXV[69]. However, its velocity of disproportionation is higher due to different substituent effects of tert-butyl and phenyl groups.

If at least one of groups R^1 or R^2 in L is alkyl or another carbon-chain group, the stability of quinone methide formed is enhanced. Several such compounds were isolated[58, 70-73]. The defined quinone methides were prepared from some important antioxidants. Thus, oxidation of methyl 3-(3,5-di-tert-butyl-4-hydroxyphenyl)propionate (Metilox®) by silver(I) oxide or lead(IV) oxide gives dimethyl 1,4-bis(4-oxo-3,5-di-tert-butyl-2,5-cyclohexadienon-1-ylidene)-2-butene-2,3-dicarboxylate LI[74]. Dimethyl 1,4-bis(4-oxo-3,5-di-tert-butyl-2,5-cyclohexadienone-1-ylidene)butan-2,3-dicarboxylate LII is a by-product, whereas dimethyl 1,4-bis(4-hydroxy-3,5-di-tert-butylphenyl)-1,3-butadiene-2,3-dicarboxylate LIII is considered as an intermediate. Similarly, octadecyl 3-(3,5-di-tert-butyl-4-hydroxyphenyl)propionate (Irganox® 1076) is oxidized. The octadecyl derivatives, analogous to LI and LII, can be isolated in a pure state only with great difficult[74].

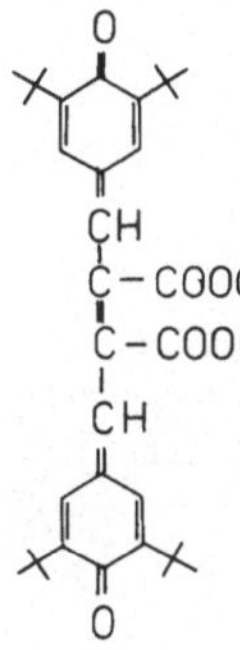

LI 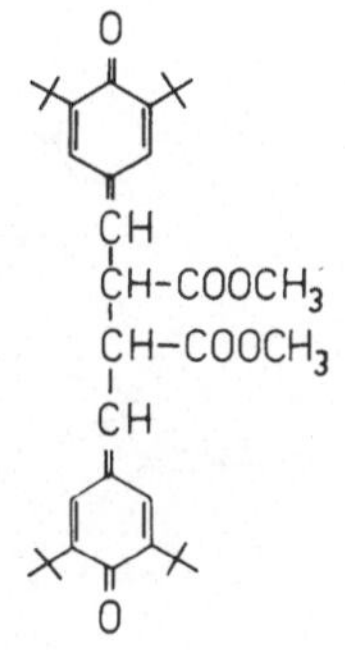   LII 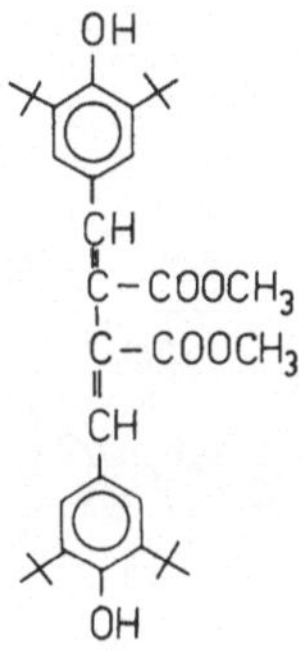   LIII

In the oxidation of Metilox®, the formation of monomeric quinone methide LIV was also detected by spectra[73]. However, it is unstable and isomerizes to LV. Also the combined quinone methide and cyclohexadienonyl derivative LVI was mentioned among reaction products.

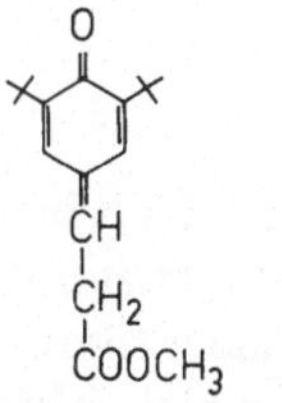

LIV 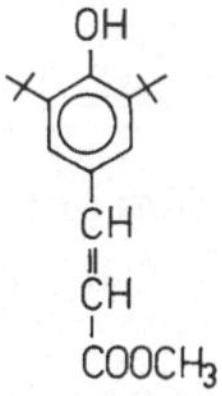   LV 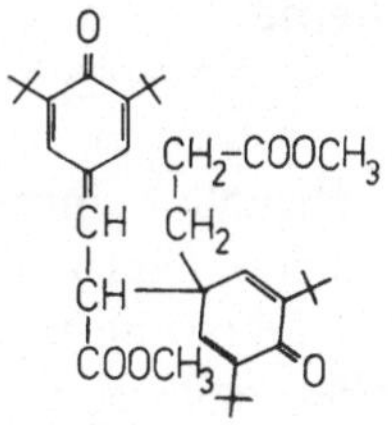   LVI

Quinone methides of the type LVII originate in oxidation of 4,4'-methylene-bis(2,6-dialkylphenols)[34, 67, 75-77]. The anion of quinone methide LVII is formed, according to[78], from **XXX** in an alkaline medium presumably by splitting off a methyl in the form of formaldehyde from the intermediary **XLIX**. Further transformations characterized by the formation of a new phenoxyl take place during the antioxidation action of these quinone methionide phenols. The compound **LVIII** („galvinoxyl") derived from 2,6-di-tert-butyl-4-(3,5-di-tert-butyl-4-hydroxybenzylidene)-2,5-cyclohexadiene-1-one (**LVII**) is among the most stable free radicals (Scheme 7). Its properties were investigated in detail by the electron spin resonance (ESR) and electron-nuclear-double resonance methods[79]. **LVIII** acts as an effective scavenger of R· and RO· radicals[80, 81]. **LIX** results in the former case and a mixture of 2,6-di-tert-butyl-4-hydroxybenzaldehyde **XXXII** and 2,6-di-tert-butyl-4-tert-butoxyphenol in the latter. **LVIII** is oxidized by molecular oxygen[82, 83] to a mixture of compounds, among which 2,6-di-tert-butyl-1,4-benzoquinone **XXII** and 3,5-di-tert-butyl-4-hydroxybenzaldehyde **XXXII** were identified[82]. According to[83], their origin is preceded by the formation of peroxy radical **LX** and a more complex quinone methionide and cyclohexadienonyl system **LXI**.

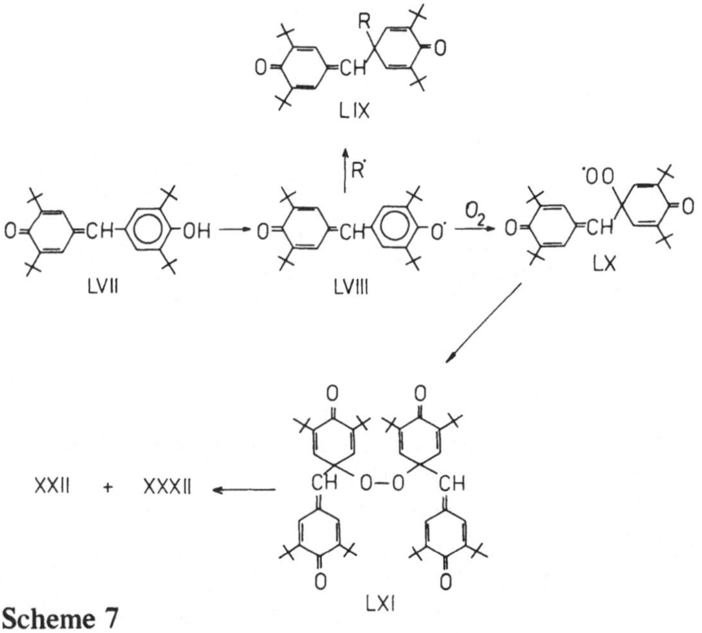

Scheme 7

Phenolic quinone methides isomeric with **LVII** have been considered as the product of oxidative transformations of another antioxidant, 2,2'-methylenebis(4-methyl-6-tert-butylphenol) **XXXVII**[84]. Rubber films, prepared from high-pH lattices stabilized with this antioxidant, turn pink during the thermal oxidation. The discolouration is ascribed[84] to the formation of ion **LXII**. However, the oxidation of **XXXVII** seems to proceed in a more complicated way. Scheme 6 shows the origin of oxidation coupling products prepared under conditions which simulated the inhibition process[49, 85]. The increased relative content of radicals tert-BuO·$_2$ in the reaction mixture increases the content of oligomeric dark-coloured coupling products. These products previal already at the molar ratio of **XXXVII** to tert-BuO˙ 1 : 2. There were identified the dimer **XLII**, trimer **XLVII**, and tetramer of the composition $C_{92}H_{118}O_8$.

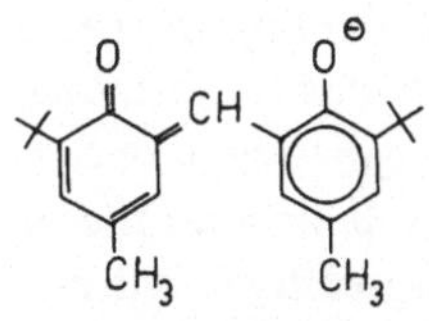

LXII

Quinone methides are also formed from further multinuclear phenols that have a hydrogen atom at the carbon atom connecting the phenolic part with the rest of molecule. In the main, only simpler structures were defined which result by disproportionation of phenoxyls. The structures of more complicated products of oxidative

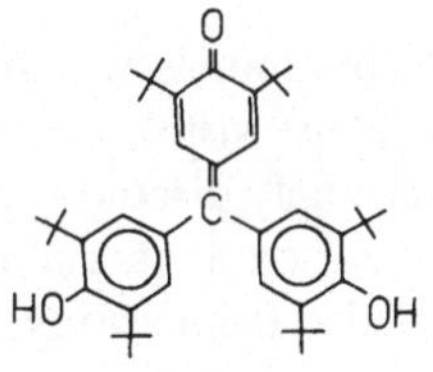

LXIII

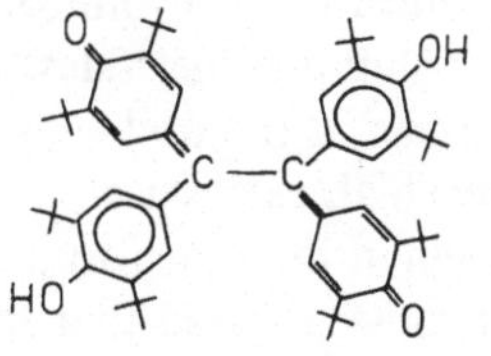

LXIV

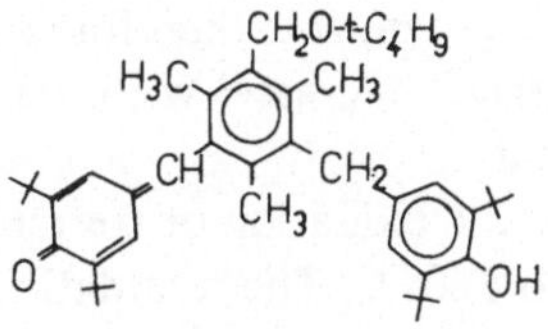

LXV

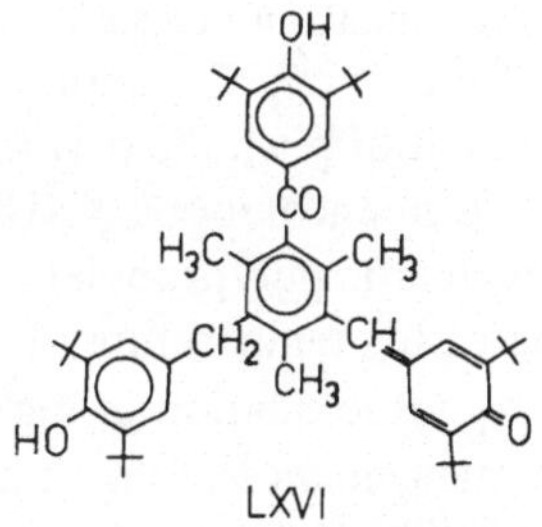

LXVI

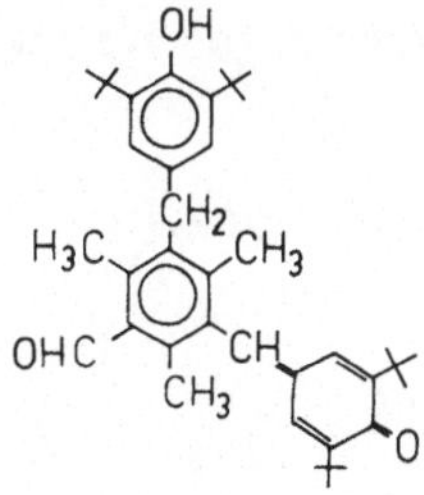

LXVII

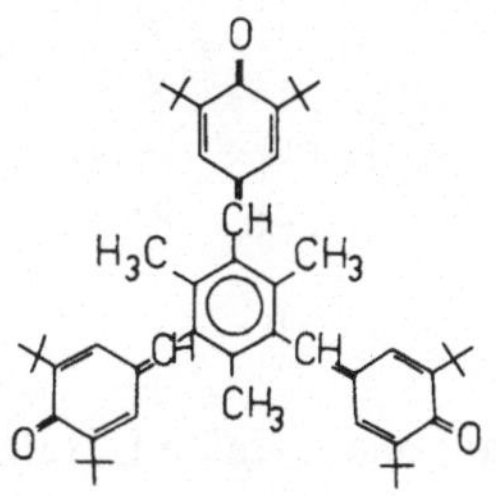

LXVIII

coupling have not been deciphered. LXIII and LXIV may serve as an example of isolated compounds[86, 87]. The structure of further compounds LXV-LXVIII, which contain, in addition to the quinone methinoid system, also further functional systems, were proposed on the basis of mass spectra for the transformation products of 1,3,5-trimethyl-1,3,5-tris(3,5-di-tert-butyl-4-hydroxyphenyl)benzene formed in the oxidation of polypropylene at 200 °C or in the oxidation of the same anti-oxidant by oxygen at this temperature[88, 89]. At the same time, the given com-pounds are examples of systems with different degrees of oxidation transformation originating in one molecule, which are possible with multinuclear phenolic anti-oxidants with the resonance effect between the individual phenolic nuclei inter-rupted by a hydrogen-containing bridge.

The compounds with the proved or presumed quinone methionide structure belong to the most frequent transformation products of phenolic antioxidants. Simpler quinone methides were identified in the oxidized hydrocarbon substrates containing stabilizers. Stilbenequinone XXIX was found in the weathered or oxida-tion-degraded polyethylene[90, 91] and in the turbine[92] or paraffin oils[91] which were stabilized with 2,6-di-tert-butyl-4-methylphenol. XLII and XLVII were found in tetrahydronapththalene and polypropylene stabilized with 2,2'-methylenebis(4-methyl-6-tert-butylphenol)[85]. The presence of compounds LVII and LXII was assumed in aged films of rubber[84].

Quinone methides, including the compounds which have further functional systems in the molecule, absorb light in the visible range of the spectrum. They have high extinction coefficients and strongly discolour polymers. However, their origin in the transformation process of antioxidants cannot be prevented. Numerous examples showed that they are formed also from antioxidants marked as nonstaining. Quinone methides are mostly very reactive[93] and may actively affect the long-term properties of polymers. The behaviour of individual quinone methides is different. If they contain also the phenolic nucleus in molecule, as it is e. g., with LVII or XLII, this part of molecule determines properties of the whole system from the standpoint of stabilization mechanism and the system acts therefore as a strong anti-oxidant[31, 55]. The antioxidation effect decreases with the increasing relative content of phenolic part in molecule.

Stilbenequinone XXIX retards the oxidation of tretrahydronaphthalene[31, 32] and atactic polypropylene[94] up to 120 °C. The retardation effect weakens with the increasing temperature of oxidation. XXIX has a strong stabilization capacity also in the photooxidation of high-impact polystyrene[95]. It undergoes the acid-catalyzed rearrangement leading to 2,2-bis(3,5-di-tert-butyl-4-hydroxyphenyl)acetaldehyde LXIX[96]. This compound may be expected to have a weak antioxidation capacity.

LXIX   LXX

Quinone methide LI retards the oxidation of tetrahydronaphthalene only at lower temperatures; a prooxidative effect was observed already at 120 °C[97]. On the other hand, it has been stated[98] that some multinuclear quinone methides, e. g., LXX, stabilize polypropylene against oxidation at temperatures as high as 150 °C. The authors interpret the efficiency of LXX by its capacity to scavenge radicals R˙. However, it seems more probable that quinone methides acts as stabilizers due to the trapping of RO˙ radicals in the sense of data in[80].

An important finding is that quinone methides may interfere with photoinduced processes which are a part of atmospheric ageing. There are only few product studies. Quinone methide XXX undergoes a photoinduced reaction in diethyl ether medium giving rise to the phenolic addition compound LXXI (R = MeCHOEt)[99]. XXIV occurs as a byproduct. LXXI (R = 2,4-cyclohexadienyl) originates in the medium of 1,3-cyclohexadiene[100].

LXXI   LXXII·   LXXIII

According to[101], LXXIII (R=H) is formed in the photolysis of diluted alcoholic solutios of quinone methide LXXII sensitized by acetophenone; however, LXXIII (where R is the residue of alcohol used: CH_2OH or $(C(CH_3)_2OH)$ originates according to[102].

Practical consequences follow from the finding that quinone methides XXIX, XLII, XLVII, or LI, which are the transformation products of technically important antioxidants, quench the singlet oxygen[74, 103]. Their quenching capacity is similar to that of β-carotene, but they are more resistant to oxidation.

It should be summarized that quinone methides, though they may discolour a polymer, do not exhibit entirely negative action from the standpoint of long-term properties of polymers because they contribute to the chain-breaking effect of antioxidants, from which they were derived, by retardation of thermal oxidation and quenching the singlet oxygen.

C. Cyclohexadienones

1. Alkylperoxycyclohexadienones

If the concentration of $RO_2^˙$ radicals is high enough, they react during the polymer stabilization with phenoxyl In˙ primarily formed according to (4) and give rise to alkylperoxocyclohexadienones (ROO–CHD). The reaction in which a resonance cyclohexadienonyl form of phenoxyl takes part, has the rate of order of mangnitude 10^8 $1.mol^{-1}.s^{-1}$ [104, 105]. The formation of ROO–CHD is in agreement with the kinetic data which reveal that monohydric phenolic antioxidants terminate two kinetic oxidation chains in the ideal case.

Former data about properties of cyclohexadienones are reviewed in[42, 43]. This chapter deals only with derivatives containing a peroxide function in the molecule, because they have a specific importance for the long-term properties of polymers. Compounds derived from phenolic antioxidants have the structure of 4-alkylperoxy-4-substituted 2,6-di-tert-butyl-2,5-cyclohexadiene-1-ones XXXVI (in Scheme 5) or 2-alkylperoxy-2-substituted 4,6-di-tert-butyl-3,5-cyclohexadiene-1-ones LXXIV, where R^1 may be alkyl, substituted alkyl, or the residue of a molecule of multi-nuclear phenolic antioxidant and R is the residue derived from the oxidized substrate.

LXXIV

ROO-CHD are not formed as the only product of the reaction of $RO_2^{\cdot}$ with phenol indeed, neither under optimum reaction conditions. The competition coupling reactions of phenoxyls described in the Chap. II A. proceed in larger extent also at the molar ratio phenol-to-$RO_2^{\cdot}$ higher than 0.5.

ROO-CHD XVI is formed in the yield of 5—10 % in the reaction of 2,4-di-tert-butylphenol with tert-$BuO_2^{\cdot}$[33] via the phenoxyl VI (Scheme 3). Its isolation from the reaction mixture is difficult, because it is cleaved on silica gel to 2-tert-butyl-1,4-benzoquinone. Also the isomeric ROO-CHD XIV is the intermediary formed from the same phenol, but it is immediately cleaved to 3,5-di-tert-butyl-1,2-benzoquinone XV. 2,2′-Biphenyldiol X, originating as a consequence of C—C coupling of phenoxyl VI, is not the final product in the presence of excess of $RO_2^{\cdot}$ and ROO-CHD's XI to XIII are formed (the compound XII in the yield as high as 60 %). Also the phenolic derivative VIII undergoes further oxidation, giving ROO-CHD IX as the final product.

The instable ROO-CHD XXI is the intermediary formed by oxidation of 2,6-di-tert-butylphenol[38]; the final product is 2,6-di-tert-butyl-1,4-benzoquinone XXII (Scheme 4).

Most information is available about compounds formed from 2,6-di-tert-butyl-4-methylphenol XXIV. This is comprehensible not only with respect to the importance of this antioxidant, but also in consideration of its relatively simple structure. The formation of ROO-CHD of the type XXXVI (R^1 = Me) was proved in the inhibited oxidation of cumene[45, 106—109], tetrahydronaphthalene[35, 55], and polypropylene[55]. The compounds with R = tert-butyl, tretrahydronaphthyl, or cumyl were prepared as models.

As regards the processes proceeding during the stabilization of polyolefins, the reaction of XXIV with various alkylperoxyls simulating the hydroperoxidized segments of polyethylene and polypropylene chain was investigated in detail. The radicals $RO_2^{\cdot}$ were generated catalytically from tert-butyl hydroperoxide LXXV, 2-methyl-2-hydroperoxyheptane, LXXVIII, 3-methyl-3-hydro-peroxyheptane

LXXIX, 2,4-dimethyl-2-hydroperoxypentane LXXVI, 2,4,6-trimethyl-2-hydro-peroxyheptane LXVII, and 2,4,6-trimethyl-4-hydroperoxyheptane LXXX.

$$CH_3-\underset{\underset{OOH}{|}}{\overset{\overset{CH_3}{|}}{C}}-(CH_2-\overset{\overset{CH_3}{|}}{CH})_n CH_3$$

LXXV: n = 0
LXXVI: n = 1
LXXVII: n = 2

$$CH_3-\underset{\underset{OOH}{|}}{\overset{\overset{CH_3}{|}}{C}}-(CH_2)_4-CH_3$$

LXXVIII

$$CH_3-\underset{\underset{OOH}{|}}{\overset{\overset{CH_2-CH_3}{|}}{C}}-(CH_2)_3-CH_3$$

LXXIX

$$\overset{CH_3}{\underset{CH_3}{>}}CH-CH_2-\underset{\underset{OOH}{|}}{\overset{\overset{CH_3}{|}}{C}}-CH_2-CH\overset{CH_3}{\underset{CH_3}{<}}$$

LXXX

4-Alkylperoxy-4-methyl-2,6-di-tert-butyl-2,5-cyclohexadiene-1-ones XXXVI,
R^1 = Me,
R = (a) t−C_4H_9
    (b) $C(CH_3)_2 C_5H_{11}$
    (c) $C(CH_3)(C_2H_5)C_4H_9$
    (d) $C(CH_3)_2 CH_2 CH(CH_3)CH_3$
    (e) $C(CH_3)_2 [CH_2 CH(CH_3)]_2 CH_3$
    (f) $C(CH_3)[CH_2 CH(CH_3)CH_3]_2$

were prepared in yields 61−80 %[110]. The course of the reaction was not negatively influenced by relatively great steric requirement of some alkyl groups, e. g. in alkyl-peroxyls derived from hydroperoxides LXXVII and LXXX. This result indicates that alkylperoxycyclohexadienones are the main reaction product at sufficient con-centration of $RO_2^\cdot$ in the system hydrocarbon polymer-phenolic antioxidant. For example, ROO-CHD XXXVIa is formed at the about twofold molar excess of tert-BuO$_2^\cdot$ with respect to XXIV in the yield of 80 %, in addition to a mixture of prod-ucts of the oxidative coupling. At the equimolar ratio of $RO_2^\cdot$ and XXIV, the com-pounds XXVIII and XXIX represent already 25 % of the reaction product. The im-portance of the relative content of $RO_2^\cdot$ was shown also on the product composition in oxidation of 2,2'-methylenebis(4-methyl-6-tert-butylphenol)[49].

The compounds LXXVI to LXXX are the more true models of oxidized poly-olefin than tert-butyl hydroperoxide. However, it followed from the results ob-tained by means of various hydroperoxides, which simulated the oxidized chains of polyolefins, that the simplest and most readily available model of polypropylene hydroperoxide − tert-butylhydroperoxide − can be used for the simulation of mechanism and transformations of phenolic antioxidants in spite of the differences in the decomposition kinetics of the polymeric hydroperoxide and its low-molecular-weight model[111].

In consequence of the intramolecular chain transfer in the oxidation of poly-olefins, more hydroperoxide groups than one is formed at one macromolecular skele-

ton[111, 176]. They are situated irregularly and most probably are in the neighbourhood of OH and > CO groups in the oxidized polymeric chain[113]. The isolated
and cumulated HOO groups differ in their properties. It can be expected that the
reaction with phenol will proceed analogously as with tertiary hydroperoxides containing only one HOO group, if HOO groups are more separated. 2,4-Dimethyl-2,4--
pentanedihydroperoxide LXXXI, which is able to react to both reaction centres, was
used as a model for the investigation of the behaviour of an oxidized segment of
polypropylene chain containing two HOO groups at the neighbouring tertiary carbon
atoms. In the course of its reaction with XXIV, 2,4-dimethyl-2,4-bis(1-methyl-3,5-
di-tert-butyl-2,5-cyclohexadiene-4-ylperoxy)-pentane LXXXII is formed[114]. In
addition to this compound, 2,4-dimethyl-2-(1-methyl-3,5-di-tert-butyl-2,5-cyclo-
hexadiene-4-onylperoxy)-4-pentanol LXXXIII arises in approximately the same
amount as a reaction product of a mixed alkoxy- and alkyperoxy radical originating
from LXXXI by the sequential reaction with phenol XXIV and the derived phenoxyl[114]. Along with LXXXII and LXXXIII, also further phenolic and quinone
methionid products are formed by the oxidation coupling, the amount of which
grows with the increasing relative content of $RO_2^{\cdot}$ in the mixture with XXIV.

Using tert-butylhydroperoxide, ROO-CHD's were prepared from various
commercially produced and technically important phenolic antioxidants. Thus,
octadecyl 3-(1-tert-butylperoxy-3,5-di-tert-butyl-2,5-cyclohexadiene-4-onyl)-
propionate LXXXIV is formed from octadecyl 3-(3,5-di-tert-butyl-4-hydroxy-
phenyl)-propionate (Irganox® 1076)[115].

LXXXIV

ROO-CHD XLIV and XLV resulted in a high yield from 2,2'-methylene-bis(4-methyl-
6-tert-butylphenol) XXXVII (Scheme 6)[49] provided that the reaction mixture
contains the sufficient amount of radicals $RO_2^{\cdot}$, which are able to transform
phenoxyls XXXVIII and XLIII to ROO-CHD before the coupling reaction or disproportionation may occur. ROO-CHD is formed also in the oxidation of other

multinuclear phenolic antioxidants. 1,3,5-Trimethyl-2,4,6-tris (3,5-di-tert-butyl-4-hydroxybenzyl)benzene (Ionox® 330) is oxidized by tert-BuO·$_2$ in benzene solution to 1,3,5-trimethyl-2,4,6-tris(1-tert-butylperoxy-3,5-di-tert-butyl-2,5-cyclohexadiene--4-onylmethyl)benzene LXXXV (R = tert-Bu)[88, 116]. The partial oxidation in tert-butyl alcohol[116] yielded 1,3,5-trimethyl-2-(3,5-di-tert-butyl-4-hydroxybenzyl)-4,6,-bis-(1-tert-butylperoxy-3,5-di-tert-butyl-2,5-cyclohexadiene-4-onylmethyl) benzene L LXXXVI.

LXXXV   LXXXVI

LXXXVII   LXXXVIII

Another important trisphenolic antioxidant for polyolefins — 1,3,5-tris-(4-hydroxy-3,5-di-tert-butylbenzyl)cyanuric acid (Good Rite® 3114) — forms by oxidation in benzene medium 1,3,5-tris(1-tert-butylperoxy-3,5-di-tert-butyl-2,5-cyclohexadiene-4-onylmethyl) cyanuric acid LXXXVII[117], while in the medium of tert-butyl-alcohol a mixture is formed which contains in addition to LXXXVII also 1-(4-hydroxy-3,5-di-tert-butylbenzyl)-3,5-bis (1-tert-butylperoxy-3,5-di-tert-butyl-2,5-cyclohexadiene-4-onylmethyl)cyanuric acid LXXXVIII as a product of partial oxidation. No compound containing two unreacted phenolic nuclei was found among the isolated products. However, a mixture of dark-coloured compounds is formed at the same time, which have according to GPC analyses molar volumes either similar to LXXXVII and LXXXVIII, or volumes corresponding to the partially degraded skeleton of the original trisphenol. It can be assumed, analogously with the transformation products of other sterically hindered phenols, that the discoloured compounds have quinone methionide structures.

The oxidation transformation of antioxidant Good Rite® 3114 is faster in comparison with the trisphenol Ionox® 330 and is probably the cause of a lower stabilization capacity of the former antioxidant rated according to the oxygen absorption in polypropylene oxidized at 180 °C[116, 117]. It seems that the cause of the different oxidation sensitivity of both trisphenols is the only different structural element, i. e. the six-membered central ring[117].

The tetra nuclear phenolic antioxidant tetrakis methylene [3-(3,5-di-tert-butyl-4-hydroxyphenyl)propionate] methane (Irganox® 1010), designed for the stabilization of polypropylene, is oxidized in the yield of about 75 % to tetrakis methylene [3-(l-tert-butylphenoxy-3,5-di-tert-butylcyclohexa-2,5-diene-4-onyl) propionate] methane LXXXIX[118].

LXXXIX   XC

All given examples of ROO-CHD's show that they have bonded in their molecule the alkyl residue of hydroperoxide through the action of which they arose. Extrapolating these results to the polymeric substrate, it should be considered the bonding of cyclohexadienonyl structures to the polymeric chain. The bonding of 2,4,6-tri-tert-butylphenol to oxidized polypropylene was proved in[119−121]. It was estimated from the UV-spectral characteristics corresponding to isomeric ROO-CHD structures XXXVI and LXXIV[42, 43, 122] that both isomers XXXVI and LXXIV (R^1 = tert-Bu, R, = polypropylene residue) were formed in the ratio 1 : 4[121].

Another proof for incorporation of the originally present antioxidant into an oxidized polypropylene was provided by ^{14}C-labelled Ionox® 330. The structures XC and LXXXV (R = polypropylene residue) were proposed for the reaction product[123]. It is, however, more probable that the transformation products of trisphenolic antioxidant are bonded to the polypropylene skeleton in a simpler way for steric reasons than under formation of a polymer network.

An unambiguous proof for the incorporation of phenolic antioxidant into a polymer as a consequence of transformations during the inhibited oxidation was obtained by means of atactic polypropylene[124]. Polypropylene hydroperoxide was prepared by its oxidation, characterized, and used for generation of peroxy radicals. Using these radicals, 2,6-di-tert-butyl-4-methylphenol was transformed. An extractable part of the obtained product was formed by a mixture of XXVIII, XXIX, and light to dark-coloured products of oxidative coupling. The other part of products was bonded to the polymeric matrix. The GPC analysis revealed that polydispersity of this product did not change in comparision with the original polypropylene hydroperoxide. The spectral investigation confirmed the presence of a structure having absorption at 232 ± 2 nm, which is characteristic of 4-alkylperoxycyclohexadienones[114]. The main evidence about the presence of structure XXXVI (R^1 = Me, R = polypropylene residue) was the formation of a radical with characteristic ESR spectrum after irradiation of polypropylene with the incorporated transformed anti-

oxidant by light of wavelength 360–480 nm. The ESR spectrum was identical with
that of radicals formed after irradiation of defined low-molecular-weight ROO-
CHD's[124, 125]. The ESR method was used also to estimate the total concentration
of ROO-CHD in the polymer chain. It was calculated from data on the stationary
concentration of radicals formed by photolysis and the content of hydroperoxide
groups in the initial polypropylene hydroperoxide that about 60 % of the originally
present HOO groups reacted with phenol XXIV[124].

The information mentioned until now concerned the formation of ROO-CHD by
the reaction of phenoxyls with radicals $RO_2^•$.

Interesting data about interactions in the system containing 2,4,6-tri-tert–butyl-
phenoxyls XCI and tert-butylhydroperoxide or tetrahydronaphthylhydroperoxide
are presented in[127] (Scheme 8). The product analysis elucidates further processes
which may occur during the inhibition action. The results show that also a sterically
hindered phenoxyl can take part in the reaction (5). Phenol XCII and $RO_2^•$ radical
are generated as a consequence. The latter immediately reacts with the remaining
phenoxyl in its resonance form XCIII giving 4-tert-butylperoxy-2,4,6-tri-tert-butyl-
2,5-cyclohexadiene-1-one XCIV as product or via the instable LXXIV ($R = R^1 =$
tert-Bu) to 3,5-di-tert-butyl-1,2-benzoquinone XV, tert-butyl alcohol and isobutylene.

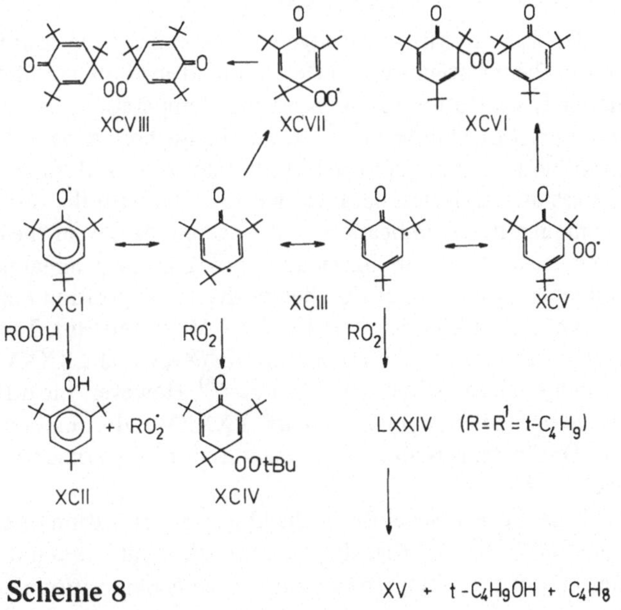

Scheme 8 $XV + t-C_4H_9OH + C_4H_8$

Owing to the presence of the peroxidic bond, ROO-CHD takes part in reactions
induced by heat and light. The peroxidic bond is split and the resulting radical frag-
ments act as initiators. The antioxidation efficiency of phenolic antioxidants may
therefore be decreased, particularly at higher temperatures, by the products of their
own transformations. For example, biscyclohexadienone XLIV, if used in the

mixture with bisphenol XXXVII from which it is derived, decreased the efficiency of XXXVII in the stabilization of polypropylene oxidized at 180 °C[55].

The decomposition temperature of various ROO-CHD's was followed by the differential scanning calorimetry (DSC). 2-tert-Butylperoxy-2-methyl-4,6-di-tert-butyl-2,5-cyclohexadiene-1-one LXXIV (R = tert-Bu, R^1 = Me) decomposes at 75 °C[31, 128]. According to[119, 121], this ROO-CHD is about thirty times less thermally stable than the analogous isomeric XXXVIa. Also[129] mentioned the lower thermal stability of 2-tert-butylperoxy-2,4,6-tri-tert-butyl-3,5-cyclo-hexadiene-1-one in comparision to 4-tert-butylperoxy-2,4,6-tri-tert-butyl-2,5-cyclo-hexadiene-1-one. In accordance with the high sensitivity to heat, the compounds of type LXXIV initiate the oxidation of tetrahydronapththalene as low as 65 °C[31, 32, 94]. It can be extrapolated from the thermal behaviour of LXXIV that 2-alkylperoxy-3,5-cyclohexadienones, formed from so called cryptophenolic anti-oxidants, accelerate the oxidation of some hydrocarbon substrates already near the ambient temperature.

The thermal decomposition of 4-alkylperoxy-4-alkyl-2,5-cyclohexadiene-1-ones XXXVI starts at temperatures 110—130 °C according to DSC measure-ments[31, 110, 114, 115]. Similar starts of decomposition were determined for ROO-CHD's derived from multinuclear phenols: 130 and 132 °C for LXXXV and LXXXVI[116], respectively, 133 and 140 °C for LXXXVII and LXXXVIII, resp.[117], and 125 °C for LXXXIX[118].

To elaborate the knowledge about thermal behaviour in the temperature region used in the investigation of effect on hydrocarbon oxidation, ROO-CHD's were iso-thermally heated for 50 min at various temperatures below the temperature at which decomposition command[128]. After the isothermal period, the heating was con-tinued with the temperature linearly increasing with time. The calculated difference of areas of exotherms, obtained without and with the isothermal heating, was used as a semiquantitative indication of the decomposition of the compounds during heating. The presence of a more bulky alkyl group in the geminal position to the ROO group facilitates the decomposition below the decomposition temperature (comparision of compounds XXXVI, R = tert-Bu, R^1 = Me or tert-Bu). The binuclear ROO-CHD XLIV and the trinuclear compounds LXXXV and LXXXVII decompose faster than the mononuclear ROO-CHD XXXVI[128]. However, the differences also occur among triscyclohexadienonyl compounds: LXXXV is decomposed faster and, in accordance with DSC data, it is also a stronger initiator in the oxidation of tetrahydronaphtha-lene[116, 117].

It has been ascertained in the kinetic investigation of thermolysis of various ROO-CHD's[119, 120] that the decomposition rate does not much depend on the character of the alkylperoxy group. The decomposition rate constants are of the same order of magnitude as those of the typical radical initiators [2,2'-azo bis(iso-butyronitrile), benzoyl peroxide] and are even higher than those of hydroperoxides, which are the main species leading to the branching of oxidation chains. This in-dicates the unfavourable properties of this type of transformation product of phenols in the oxidation of polymers and hydrocarbons, ROO-CHD XXXVIb, LXXIV, LXXXIV, LXXXV or XCVIII initiate, for example, the oxidation of tetrahydro-naphthalene at 100—120 °C more effectively than α, α-dicumyl peroxide[31].

The DSC and thermogravimetric analytical (TGA) information about thermal properties of polypropylene cyclohexadienone XXXVI (R^1 = Me, R = polypropylene residue) is interesting[124]. The DSC curve exhibits an exothermic double peak which begins at 98 °C and has maxima at 137 and 167 °C. Comparsion with the DSC decomposition curve of the initial polypropylene hydroperoxide allowed the peak at 137 °C to be assign to the cyclohexadienonyl structure bonded to the polypropylene skeleton. The decomposition of polymeric peroxycyclohexadienonyl derivative begins at a lower temperature than that of its low-molecular-weight analogues. It is difficult to estimate the extent of affection of its decomposition temperature by the vicinity of other functional groups in the chain. The steric factor of polymeric chain plays presumably a more important role.

It is striking how little information exists about the product of ROO-CHD thermolysis, though this reaction is of great importance from the point of view of long-term properties of polymers. The resulting mixtures of compounds have their composition considerably influenced, even by experimental conditions. The mixture formed by decomposition of LXXIV (R = tert-Bu, R^1 = Me) contained, in addition to others, also a compound which behaved exothermally in the thermal analysis[128]. According to[31], XXXVI (R = tert-Bu, R^1 = Me) is thermolyzed in melt under formation of 3,5-di-tert-butyl-4-hydroxybenzaldehyde XXXII (24 %), 2,6-di-tert-butyl-1,4-benzoquinone XXII (~ 15 %), stilbenequinone XXIX (10 %) (determined by GPC), and of two unidentified compounds, in addition to the traces of XXIV. If the same compound was decomposed in xylene, a mixture of products was formed, containing a compound of peroxidic character which originated by interaction between ROO-CHD and the hydrocarbon solvent, 2,6-di-tert-butyl-1,4-benzoquinone XXII, 3,5-di-tert-butyl-4-hydroxybenzaldehyde XXXII (37 % of the product according to GPC), and stilbenequinone XXIX[130]. In agreement with this result is the fact[91, 92] that 3,5-di-tert-butyl-4-hydroxybenzaldehyde XXXII was identified in a turbine oil oxidized at 165.5 °C in the presence of 0.6 % XXIV or in polyethylene or decane[91] oxidized in the presence of the same antioxidant at 120–200 °C.

Acetone, which is considered as product of further splitting of the primarily liberated tert-butoxyl, was identified on heating of XXXVI (R = tert-Bu, R^1 = Me)[131].

According to[121], 2,6-diphenyl-4-tert-butylperoxy-4-methoxy-2,5-cyclohexadien-1-one is decomposed under formation of methyl alcohol and 2,6-diphenyl-1,4-benzoquinone. 4-tert-Butylperoxy-2,4,6-tri-tert-butyl-2,5-cyclohexadiene-1-one XXXVI (R = R^1 = tert-Bu) is thermolyzed already at 65 °C to the mixture of compounds, where 2,6-di-tert-butyl-1,4-benzoquinone XXII was identified as the main component[31, 130]; the mixture contained, according to GPC, also 3,5-di-tert-butyl-4-hydroxybenzaldehyde XXXII and stilbenequinone XXIX.

Naturally, information about the products of thermolysis of multinuclear ROO-CHD's are even more scarce. A mixture of 10 compounds arose from LXXXV at 170 °C[116]. Some of them were able to reduce the ferro-ferricyanide agent for phenols and some had p-benzoquinone character according to IR spectral analyses. Also the compounds with aldehyde functions described in[89], e. g., LXVII, may be considered as thermolysis products of LXXXV.

The initiation effect of 4-alkylperoxy-2,5-cyclohexadienones in the thermal oxidation of tetrahydronapththalene[31, 32, 110, 114–116, 118], cumene[104], cetane[94], atactic[94] and isotactic polypropylene[104, 116, 120] is in agreement with their thermal cleavage. They begin to initiate at temperatures near 100 °C[31, 32].

It followed from the investigation of initiation of tetrahydronaphthalene oxidation by model ROO-CHD's XXXVIa–f at 120 °C[110, 115] that they act above all in the early stage of oxidation. Although the effect of alkyl in the ROO group was not very impressive, it was possible to extrapolate from the results in this phase of oxidation that the compounds XXXVIb and XXXVIc, which simulated the bonding of cyclohexadienonyl group to a polyethylene chain, initiated most weakly. The compounds XXXVId-f simulating a polypropylene chain were more efficient. Bis-peroxycyclohexadienone LXXXII initiates in the same way as XXXVIa[114] at the comparably same molar concentration of ROO groups at 120,°C. It seems, that an inductive effect of two peroxycyclohexadienonic groups present in one molecule does not occur.

The products of oxidation transformations of the trisphenolic antioxidant Good Rite® 3114 – LXXXVIII, and particularly LXXXVII – and the tetranuclear alkyl-peroxycyclohexadienone LXXXIX[118] rather negatively affect the stability of tetra-hydronapthalene in oxidation as low as at 65 °C. Both compounds LXXXVII and LXXXVIII strongly accelerate the oxidation of polypropylene at 180 °C[117].

If the transformation product of a multinuclear phenolic antioxidant contains in addition to the alkylperoxycyclohexadienone structure also the unchanged phenolic nucleus, it can possess weak antioxidative properties at temperatures when the thermolysis of peroxide does not play an important role. For example, LXXXVI weakly stabilizes tetrahydronaphthalene still at 65 °C. However, the stabilizing effect is strongly decreased in comparison to the initial antioxidant[116]. At 120 °C, the same compound affects the oxidation in the same way as the triscyclohexadienonyl derivative LXXXV.

The absorption of oxygen in the presence of ROO–CHD has a very characteristic course. The oxidation velocity retards after the rapid initial phase. Judging from the known characteristics of products of the thermal transformation of ROO-CHD, some of them having the benzoquinone character, this retardation of oxidation can be ascribed just to the products of subsequent transformations which gradually accumulate in the oxidized mixture. This explanation is in agreement with the experimentally found retardation effect of the mixture of thermolytic products of LXXXV, which was independently prepared and added to oxidized tetrahydronaphtalene. The induction period of oxidation was very short also in polypropylene which contained LXXXV or LXXXVI at 180 °C, but it was followed by a phase with the perceptible weak retardation effect of thermolysis products[116].

The weakened initiation effect of ROO–CHD, after accumulation of the products of their thermolysis which act as retarders, was shown in the oxidation of slowly oxidized substrates, e. g. cetane (a model of polyethylene) and atactic poly-propylene[94], where their accumulation is demonstrated very impressively.

Solar radiation plays an important role in weathering of polymers. This fact even more complicates the transformation of antioxidants. Because ROO–CHD's contain both a peroxidic bond and a carbonyl group in one molecule, they take part in the light-induced reactions. Isolation of products indicating the n-π^* and π-π^* excitations was reported in the photoinduced reactions of substituted 2,5-cyclohexadienones which do not contain the ROO-group[132]. The more complex transformations may be expected in ROO—CHD, because the peroxidic bon is easily photo-dissociated[133].

The facile photochemical transformation of ROO-CHD's is in agreement with their UV spectrum. They exhibit an intense band with the maximum at 233 nm (log ϵ = 4.00) and a weaker diffuse band 324—430 nm with the maximum at 375 nm (log ϵ = 1.36). The energy of absorbed light corresponds to 319.8 kJ/mol at 375 nm and suffices for the cleavage of RO-OR^1 bond, which has the strength 150.7—175.8 kJ/mol depending on the character of substitution[133]. The photochemical stability decreases with the increasing number of ROO-CHD systems in one molecule.

XXXVIa is photolyzed in a benzene or heptane solution to a radical, which is characterized by the doublet spectrum[125, 134] and which arises in heptane on irradiation by 400—480 nm light already at −65 °C [125]. The intensity of the ESR signal increased with rising temperature. The optimum conditions for measurement were achieved at 10 °C. Further increase of temperature led to the decrease of intensity of the primarily formed radical and a new spectrum was recorded. It is not certain if this secondary spectrum belongs to only one radical of if it results from superimposition position of more radicals. The fast decay of radical intermediates occurred at above 50 °C or after the irradiation had been interrupted. The structure and scheme of the formation of primary radical C was proposed by[134, 141].

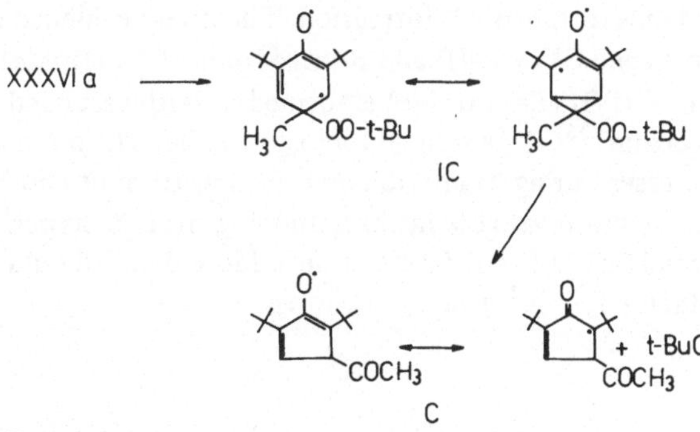

Scheme 9

The velocity of decrease of content of XXXVIa was followed in irradiation of heptane solutions by light of wavelength 333 and 463 nm in an inert atmosphere[125]. The first-order velocity of photolysis of XXXVIa was observed at both wavelengths. A mixture of compounds with the gradually changing composition is formed during the reaction. Irradiation of the benzene solution of XXXVIa, which was carried out to the conversion of 20%, gave a mixture of compounds; CI—CV (yields in %) were isolated from this mixture by means of thick layer chromatography[134].

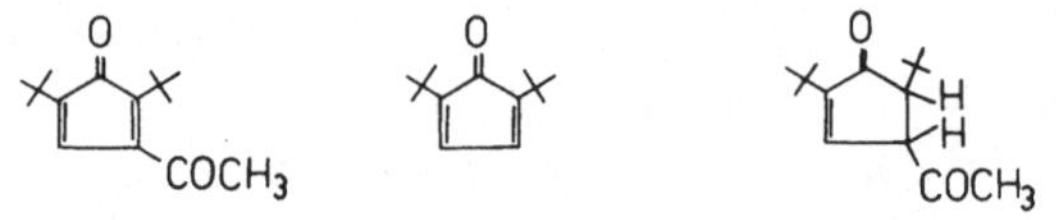

(14%)   (3%)   (12%)
CI   CII   CIII

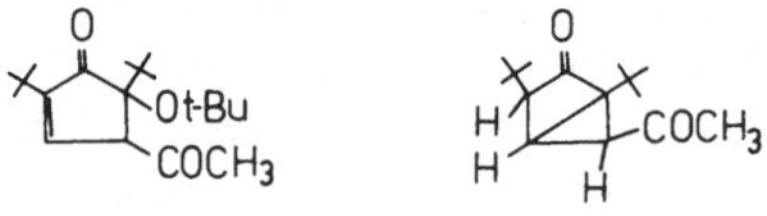

(12%)   (21%)
CIV   CV

The photolysis of solid **XXXVIa** under irradiation by light of wavelength above 400 nm proceeds more simply[125] and 3-acetyl-2,5-di-tert-butyl-2,4-cyclopenta-diene-l-one CI ist the main transformation product.

The tert-butoxy radical is liberated by photolysis of **XXXVIa** (Scheme 9). Its formation was confirmed by reactions which produce more stable radicals detectable by the ESR method[135]. The heptane solution of **XXXVIa** was irradiated by light 360—480 nm in the presence of 2,6-di-tert-butyl-4-methylphenol or 2,4,6-tri-tert-butylphenol. These sterically hindered phenols were transformed by the liberated tert-butoxyl into phenoxyls. Their identification by the ESR method is the indirect evidence of tert-BuO' formation. The direct evidence was achieved by means of nitrosobenzene **CVI** (R=H) and nitrosodurene **CVI** (R=Me) as spin-traps. The nitroxide radical **CVII** (R=H or Me) is formed in both cases and was identified by ESR spectrum[135]. This spin-adduct is unstable. The intensity of its spectrum gradually decreases during irradiation and the spectrum of another radical appears which is less intense but invariable in time. Judging from its hyperfine structure, it may be formulated as **CVIII** (R=H or Me). More detailed data on the character of substituents R^1−R^2 is not yet known.

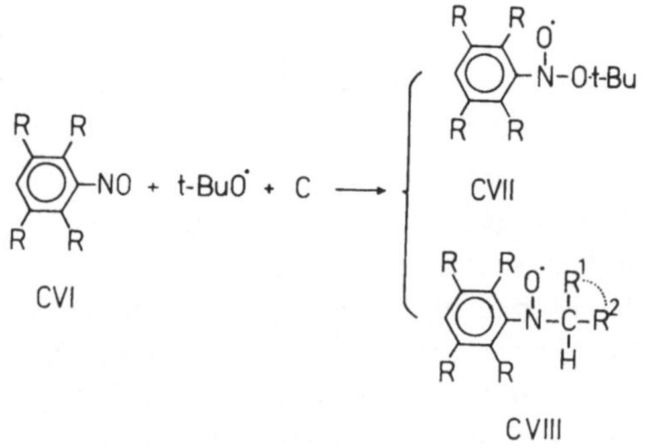

Scheme 10

The investigation[135] confirmed the cleavage of peroxidic bond of the
ROO-CHD type XXXVI in the photolysis by light of wavelength range taking part in
the weathering of stabilizied polymers. The results of photolysis were the same if the
reaction was carried out in air or in an inert atmosphere. The cleavage of ROO-CHD
occurred even in the presence of naphthalene, which is an efficient quencher of excited carbonyl groups.

A similar course of photolysis may be expected also with other ROO-CHD's. For
example, both the radical intermediate and some of the reaction products were
identified in the photolysis of 2-tert-butylperoxy-2-methyl-4,6-di-tert-butyl-3,5-
cyclohexadiene-l-one LXXIV (R = tert-Bu, R^1 = Me)[136].

It is obvious that ROO-CHD's formed by transformation of phenolic antioxidants play an active role in the atmospheric ageing of polymers. The formation of a polymeric PPO˙ radical in photolysis follows from the identity of spectra of
the radical originating by photolysis of low-molecular-weight ROO-CHD of type
XXXVI and its polymeric analogue[124]. Polypropyleneoxyl PPO˙ reduces the effect
of phenolic antioxidants by transforming them to phenoxyls, an ineffective loss of
stabilization capacity. The PPO˙ also participates in the transfer of oxidation chain
and the degradation of polymer chain takes place as the consequence of β-cleavage.
Consequently, the photolysis of ROO-CHD is negatively manifested in more aspects
of long-term properties of polymers[137].

The initiation effect of ROO-CHD was experimentally shown in the acceleration
of photo-oxidation of heptane[138], 1,3,5-tri-methylcyclohexane[139], and high-
impact polystyrene[140].

According to[142], also the catalytic cleavage of ROO-CHD and formation of
RO˙ radicals in this way have to be concerned among the processes which play a
role in the weathering of polymers.

2. Alkoxycyclohexadienones

Radicals RO˙ play an important role in the oxidation mechanism of hydrocarbons
and carbon-chain polymers. It was proved that phenoxyls are formed from phenols
by their action[44, 49, 85, 143, 144] followed by formation of phenolic and quinone
methionide compounds in coupling reaction and by disproportionation. Also the reaction between RO˙ and In˙ and formation of alkoxycyclohexadienones of the type
CIX may be presumed. Such compounds have not been isolated from the reaction
mixtures yet. However, their formation was proved by spectra[145, 179] and they
probably appear only temporarily during the inhibited oxidation due to their low
stability. According to[144], the following equilibrium reaction takes place:

$$\tag{7}$$

CIX

3. Dioxycyclohexadienones

The 2,2'- or 4,4'-dioxycyclohexadienones of type XCVI or XCVII (Scheme 8) are the products of oxidation of 2,4,6-trialkylphenoxyls by molecular oxygen[26, 58, 61, 146, 147]. The reaction proceeds in two steps. Cyclohexadienonylperoxyls, e. g., XCV or XCVII are formed first[67]. In the second step, the coupling of cyclohexadienone nuclei occurs in the position 4,4'-[58, 61, 146] or 2,2'-[26, 147] according to the character of substituents. Also the coupling 2,4'- is possible.

The bonding velocity of oxygen depends on the reactivity of phenoxyls[82, 147]. Phenoxyls having a structure which allows an extensive delocalization of unpaired electron, e. g., 2,4,6-triphenylphenoxyl[148], are extremely stable to oxygen. The more reactive sterically hindered phenoxyls, e. g., 2,4,6-tri-tert-butylphenoxyl, bind oxygen quantitatively and this reaction may be used even for analytical purposes[177].

Catalytically active metallic ions may play an important role in ageing of polymers. Their effect may become evident also in transformations of an antioxidation system. From this standpoint, data[149] dealing with oxidative transformation of 2,6-di-tert-butyl-4-alkylphenols in the presence of Co(II) complexes of Schiff bases are interesting. The mixture of compounds which results in this case contains 2,6,2',6'-tetra-tert-butyl-4,4'-dioxycyclohexadiene-1-one XCVIII and compounds formed by its cleavage, among them 2,6-di-tert-butyl-4-hydroxy-4-alkyl-2,5-cyclohexadiene-1-one of the type CX, 2,6-di-tert-butyl-1,4-benzoquinone XXII, and 3,5,3',5'-tetra-tert-butyl-4,4'-diphenoquinone XX.

CX

The formation of XCVIII was explained in[150] as a product of interaction between 2,4,6-tri-tert-butylphenol CXI and polypropylene hydroperoxide during the inhibited oxidation of polypropylene at 170 °C. The proposed process (Scheme 11) consists of the successive formation of polypropyleneoxy radicals PPO', cyclohexadienonyloxyl CXI, and 4,4'-dioxycyclohexadienone XCVIII; i. e., the same compound which is commonly formed in the reaction of phenoxyls with oxygen. It cannot be excluded that at least a part of XCVIII originates by this reaction, even under the described conditions.

4,4'-Dicyclohexadienones are relatively unstable and undergo thermo- and photoinduced reactions. Their formation in the course of inhibited oxidation of polymers may be considered as transient. The compounds formed by their successive transformations may be more likely identified in the stabilized substrate. 4,4'-Dioxycyclohexadienones are cleaved already at temperatures below 100 °C[128] under

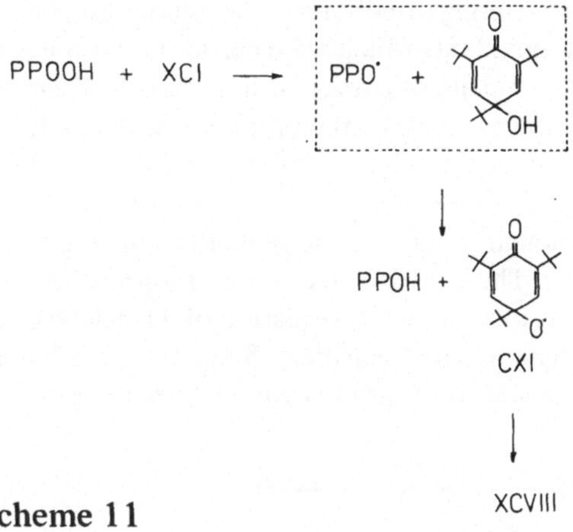

$$PPOOH + XCl \longrightarrow \boxed{PPO^{\cdot} + \text{[structure: OH quinone]}}$$

$$\downarrow$$

$$PPOH + \text{[structure: quinone-O}^{\cdot}\text{]}$$

CXI

$$\downarrow$$

Scheme 11   XCVIII

a homolysis of both O-O and C-C bonds. For example, the thermolysis of XCVIII[31, 146, 151, 152] gave a mixture of 2,6-di-tert-butyl-1,4-benzoquinone XXII, 2,6-di-tert-butyl-4-tert-butoxyphenol CXII, isobutane, and isobutylene (Scheme 12).

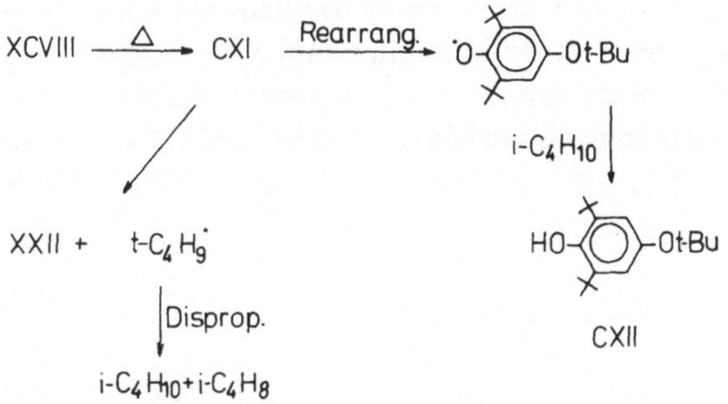

$$XCVIII \xrightarrow{\Delta} CXI \xrightarrow{Rearrang.} \text{'}O\text{-}\langle\ \rangle\text{-}O\text{-}t\text{-}Bu$$

$$\swarrow \qquad\qquad i\text{-}C_4H_{10} \downarrow$$

$$XXII + \quad t\text{-}C_4H_9^{\cdot} \qquad\qquad HO\text{-}\langle\ \rangle\text{-}O\text{-}t\text{-}Bu$$

$$\downarrow Disprop. \qquad\qquad CXII$$

$$i\text{-}C_4H_{10} + i\text{-}C_4H_8$$

Scheme 12

4. Hydroperoxycyclohexadienones

They are reasons leading to the conclusion that the singlet oxygen[133, 153] is involved in some sensitized oxidations. $^1O_2$ is able to penetrate to a considerable depth under the polymer surface[154]. Knowledge of the reactions of phenolic antioxidants with singlet oxygen is important from the view of complete understanding of processes taking part in the atmospheric ageing of stabilized polymers. It is obvious from the literature that both the quenching of singlet oxygen and the oxidation reactions occur. The quenching by phenolic compounds have been recently reviewed in[155]. They were determined rate constants for reactions of some sterically hindered phenols with $^1O_2$ [156]. An electron transfer from phenol to $^1O_2$ was assumed to be the primary step.

Product studies are of high value. Several studies were published concerning the sensitized photooxidation of phenols. With respect to the processes taking part in weathering of polymers, the studies on phenols having structures similar to effective antioxidants which were carrid out with irradiation with wavelength higher than 290 nm are of particular importance. These studies are complicated by the fact than some products of the primary transformation have limited stability under the conditions of irradiation. It is therefore advantageous to use coloured sensitizers and to irradiate the reaction mixtures by visible light in the wavelength region providing still enough energy for the generation of singlet oxygen.

The sensitized oxidation is initiated by the absorption of a light quantum by a sensitizer S under formation of a singlet-excited sensitizer which is transformed to a triplet-excited sensitizer. Singlet oxygen $^1O_2$ is formed in the system by the energy transfer to the ground state triplet oxygen[157–159].

$$^1S \xrightarrow{\ h\nu\ } {}^1S* \longrightarrow {}^3S* \tag{8}$$

$$^3S* + {}^3O_2 \longrightarrow {}^1S + {}^1O_2 \tag{9}$$

The reactions which can occur with sterically hindered phenols in the presence of $^1O_2$ and a sensitizer are shown on the example of 2,6-di-tert-butyl-4-methyl-phenol XXIV in Scheme 13. It may be expected that the oxidation of XXIV by the excited sensitizer proceeds to phenoxyl XXV, or that phenol XXIV is attacked by $^1O_2$ giving either again phenoxyl XXV or, by the 1,4-addition, endoperoxide CXIII, which is transferred to hydroperoxycyclohexadienone CXIV[160]. The latter compound may also originate via biradical CXV, as stated in[161] (Scheme 13).

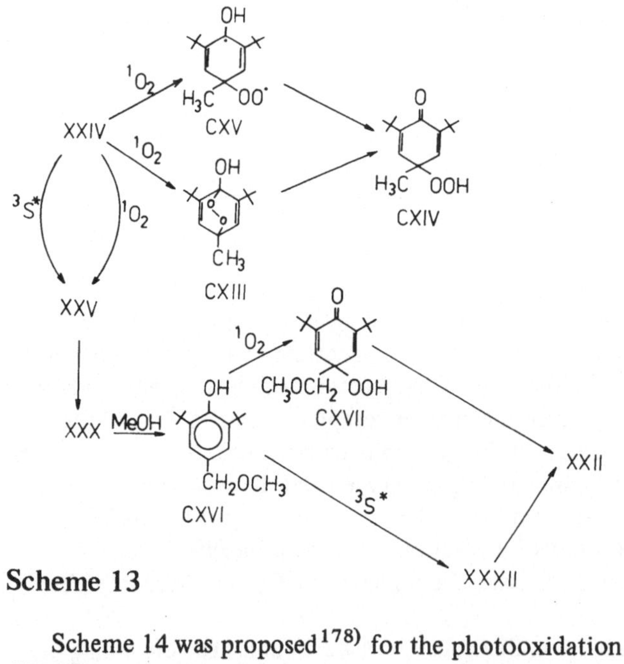

Scheme 13

Scheme 14 was proposed[178] for the photooxidation of 2,6-di-tert-butylphenol CXVIII sensitized by Bengal Red: 4,4'-biphenyldiol XIX and further diphenoquinone

XX are formed in the process via phenoxyl XVII and the C–C coupling, or 4-hydro-peroxycyclohexadienone CXXI, or 4,4'-dioxycyclohexadienone CXXII, which are photolyzed to 2,6-di-tert-butyl-1,4-benzoquinone XXII, arises via cyclohexadienonyl CXIX and cyclohexadienonylperoxyl CXX (Scheme 14).

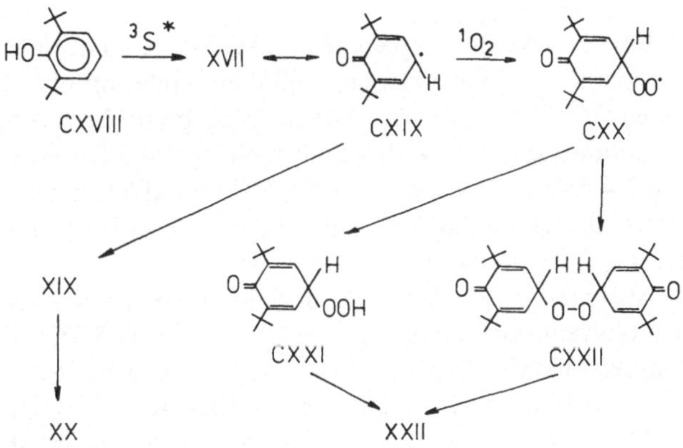

Scheme 14

2,6-Di-tert-butyl-4-methylphenol XXIV is still one of the most important anti-oxidants for polyolefins. Its structure is present also in further types of commercially produced non-volatile phenolic antioxidants. 2,6-Di-tert-butyl-4-hydroperoxy-4-methyl-2,5-cyclohexadienone CXIV (in the yield of 2.8%) and 4-methoxymethyl-2,6-di-tert-butylphenol CXVI (yield 4.2%) arose in the oxidation sensitized by Eosin Y under irradiation by a tungsten lamp in methyl alcohol medium[160] (Scheme 13). If Methylene Blue was used as a sensitizer absorbing light of longer wavelength, it was possible to carry out the oxidation in methanol under milder conditions. A rapid absorption of oxygen was observed[162] at low concentrations of phenol and the sensitizer (5 and 0.1 mol/l, resp.), which virtually stopped at the consumption of 1 mol O_2 per mol XXIV. Under the conditions used, CXIV was rather stable and was isolated in the yield of 72%. If the concentration of XXIV or sensitizer in the re-action mixture was increased, the process mechanism changed: quenching of $^1O_2$ or $^3S*$ occurred and, consequently, the composition of product changed because CXVI resulted in addition to CXIV. The formation of hydroperoxide CXIV is completely suppressed at high concentrations of phenol and sensitizer (100–130 and 2–5 mol/l, resp.). Experimental data[162] indicate that phenoxyl is more likely formed via the excited sensitizer than by oxidation with singlet oxygen. On the other hand, the latter agent plays an important role in the formation of hydroperoxycyclohexa-dienone CXIV. The photooxidation mechanism of XXIV was confirmed by the model oxidation with chemically generated $^1O_2$ and by co-oxidation of XXIV and 2,3-dimethyl-2-butene[162].

The compound CXVI is further changed during the process (Scheme 13). The oxidation by $^1O_2$ leads to hydroperoxide CXVII, while 3,5-di-tert-butyl-4-hydroxy-benzaldehyde XXXII and 2,6-di-tert-butyl-1,4-benzoquinone XXII arise by the action of $^3S*$.

Phenol XXIV is also oxidized to CXIV in the presence of Methylene Blue in dichloromethane solution[103], but the reaction proceeds more slowly than in methyl alcohol. It is remarkable that 2,4,6-tri-tert-butylphenol is reported in[163] only to quench $^1O_2$ without a substantial chemical reaction. This different behaviour in comparison to phenol XXIV may probably be explained by different experimental conditions used.

For the sake of completeness, it should be mentioned that the formation of hydroperoxycyclohexadienone CXIV by oxidation of XXIV was described[34, 164–166], as well as of the mixture of 2-hydroperoxy-2,4,6-tri-tert-butyl-3,5-cyclohexadienone with 4-hydroperoxy-2,4,6-tri-tert-butyl-2,5-cyclohexadienone from 2,4,6-tri-tert-butylphenol[166, 167] by oxidation with the triplet ground state oxygen in alkaline medium. This process proceeds by another mechanism than that described in Scheme 13.

Also methyl 3-(3,5-di-tert-butyl-4-hydroxyphenyl)propionate (Metilox®) and the important non-volatile antioxidant octadecyl 3-(3,5-di-tert-butyl-4-hydroxyphenyl)propionate (Irganox® 1076) are photooxidized[74] in the process sensitized by Methylene Blue to hydroperoxides CXXIII (R = Me or $C_{18}H_{37}$) in the yield of 57 and 38%, respectively. Their oxidation is slower than that of XXIV. It is, however, also retarded after absorption of 1 mol O_2 per mol of antioxidant.

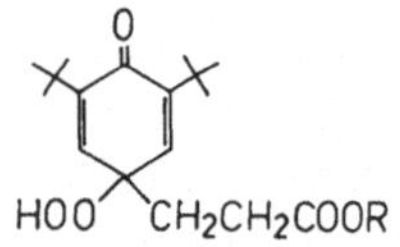

CXXIII

In accordance with[168], α-tocopherol CXXIV which is a biological antioxidant acts by a combination of chemical reaction and physical quenching. The relatively unstable hydroperoxycyclohexadienone CXXV is formed first in its reaction with $^1O_2$ and the final reaction products α-tocopheroquinone CXXVI and α-tocopheroquinonepoxide CXXVII arise by its further transformation.

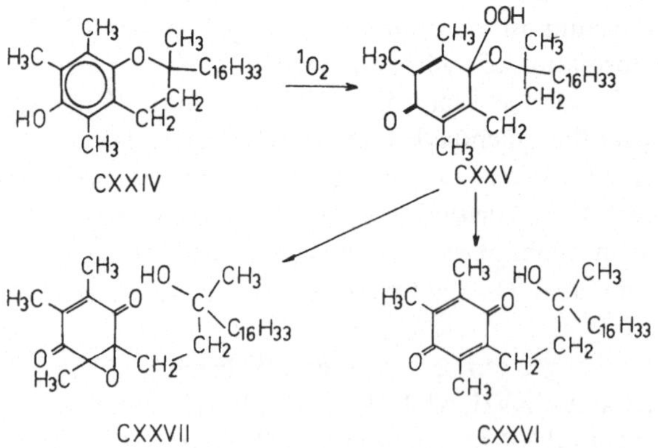

Scheme 15

According to [169], 2,2'-methylenebis(4-methyl-6-tert-butyl-phenol) XXXVII acts as an quencher in the Norrish cleavage of ketones, which is one of important processes accompanying the photooxidation of polyolefins. Some products of the transformation of antioxidant XXXVII probably also take part in this reaction, similarly as it is in quenching of $^1O_2$ by quinone methionide compounds of the type XLII or XLVII[103].

The photooxidation of XXXVII sensitized by Methylene Blue is affected by the same factors as the photooxidation of XXIV. At a low concentration of the sensitizer (0.001 mol/l), 2-tert-butyl-4-hydroperoxy-4-methyl-6-(2-hydroxy-5-methyl-3-tert-butylbenzyl)-2,5-cyclohexadiene-l-one CXXVIII arises by oxidation with the singlet oxygen[170] and is further oxidized to the mixture of diastereomeric hydroperoxides CXXIXa, b. A small amount of 2-tert-butyl-4-hydroperoxy-4-methyl-6-(2-hydroxy-5-methoxy-methyl-3-tert-butylbenzyl-2,5-cyclohexadiene-l-one CXXX is formed at the same time. The oxidation almost stops after 2 mol O_2 per mol XXXVII has been consumed. The formation of hydroperoxycyclohexadienones is suppressed by increasing the concentration of Methylene Blue to 0.2 mol/l and 2-tert-butyl-4-methyl-6-(2-hydroxy-3-tert-butyl-5-methoxymethylbenzyl)phenol CXXXI and 2,2'-methylenebis(4-methoxymethyl-6-tert-butylphenol) CXXXII, in addition to further products, are formed in methanolic medium in the presence of $^3S*$ via phenoxyl without participation of $^1O_2$.

CXXVIII   CXXIX a,b

CXXX   CXXXI   CXXXII

The product studies of transformations of phenolic antioxidants by the singlet oxygen elucidate the role of this process during weathering of polyolefins. The results of model studies[160, 162] are in agreement with the processes occurring in a real system: The presence of hydroperoxycyclohexadiene CXIV in a mixture with other compounds was proved by highperformance liquid chromatography (HPLC) in the extract of aged polyethylene which was stabilized by 2,6-di-tert-butyl-4-methyl-phenol XXIV[171].

In spite of high preparative yields, hydroperoxycyclohexadienones are not quite stable products of transformations of phenolic antioxidants being further changed thermally, photochemically, and catalytically. The compound CXIV decomposes to high extent already at temperatures below 100 °C [128] under formation of XXIV

and 4-hydroxy-4-methyl-2,6-di-tert-butyl-2,5-cyclohexadiene-l-one CX[162, 165].
3,5-Di-tert-butyl-4-hydroxybenzaldehyde XXXII was identified among further products[162].

The formation of phenol, from which thermolysed hydroperoxycyclohexadienone is derived, and of the corresponding hydroxycyclohexadienone was observed also in the thermolysis of CXXIII (R = Me or $C_{18}H_{37}$) at 130 °C[74] or of CXXVIII and CXXIXa,b (the decomposition of the last mentioned compounds begins at about 100 °C and about 90 °C, resp.[170]). The hydroperoxide CXVII is very labile and was thermolyzed[162] to 2,6-di-tert-butyl-1,4-benzoquinone XXII already at ambient temperature.

The knowledge of thermolytic products of hydroperoxycyclohexadienones explains their experimentally determined effect on the oxidation of tetrahydronaphthalene: CXIV retards the oxidation at 65 °C and retains this property also at temperatures above 100 °C[31, 32]. The low concentration of XXIV which is formed from CXIV by gradual thermolysis is responsible for this effect. The hydroxycyclohexadienone CX has only a very weak retardation effect at 65 °C and does not affect the oxidation process at all at temperatures above 100 °C[31, 32].

Hydroperoxycyclohexadienones CXIV and CXXIII are relatively stable against light having wavelengths higher than 450 nm[74, 162]. Photolysis or photooxidation take place at lower wavelengths. The photooxidation of CXIV by oxygen in heptane solution was followed in detail at the wavelengths higher than 355 nm, when the HOO-group is resistant to photolysis. The observed transformations were caused by a weak absorption of light exhibited by CXIV at 376 nm. It was found[172] that unstable 4-acetyl-2,5-di-tert-butyl-2-hydroperoxy-3-cyclopentene-l-one CXXXV is formed during the reaction. Its concentration in the reaction mixture reaches a maximum after absorption of about l mol O_2 per mol CXIV. Then, the photooxidation slows down. After absorption of 2.1 mol O_2/mol CXIV, the reaction mixture contains only traces of CXXXV. The Scheme 16 was proposed for the mechanism of CXXXV formation. The reaction mechanism is in the first phase similar to di-π-methane rearrangement[173]. However, the rearrangement is not completed owing to the presence of oxygen and a rapid trapping of oxygen by the intermediate occurs giving rise to the peroxy radical CXXXIII. A hydrocarbon substrate serves as a hydro-

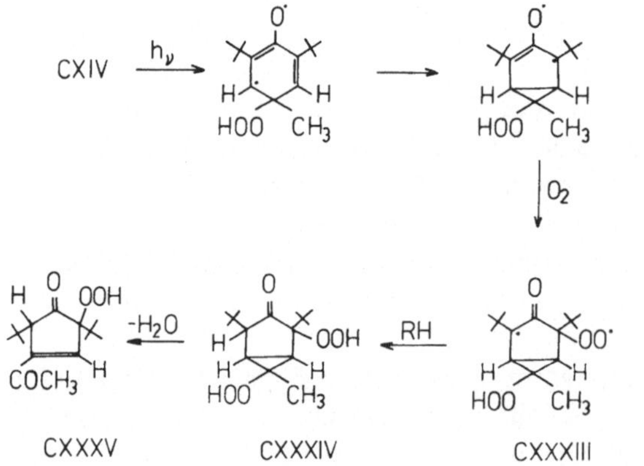

Scheme 16

gen donor in the transformation of CXXXIII to hydroperoxide CXXXIV. The reaction proceeds in the same way in heptane and in hexadecane and may be assumed
also in polyolefins. The interpretation of mechanism is in agreement with observations[103, 172] that the photooxidation of CXIV proceeds faster than its photolysis.

Hydroxycyclohexadienones of the type CX take part in photoinduced reactions
which yield a mixture of compounds containing 4—6 membered rings[174, 175].
According to our present knowledge, this process does not significantly influence
the long-term properties of polymers.

For the sake of complete knowledge of transformations, it should be mentioned
that hydroperoxycyclohexadienone CXIV is already catalytically decomposed by
Co[++] ions at ambient temperature. In the presence of XXIV, it yields a mixture of
compounds comprising hydroxycyclohexadienone CX, ethylenebisphenol XXVIII,
and stilbenequinone XXIX[112].

III. Dihydric Phenols

Alkylated hydroquinones and particularly pyrocatechols stabilize polypropylene[180, 181] and low-molecular weight hydrocarbons, e. g. tetrahydronaphthalene
well against oxidation[182]. Also more complicated derivatives of pyrocatechol are
very effective antioxidants in polypropylene, as e. g. 3,3,3',3'tetramethyl-1,1'spiro-
biindane-5,6,5',6'-tetrol or 3,3,3',3'-tetramethyl-1,1'-spirobiindane-6,7,6',7'-
tetrol[183]. However, the disadvantage of these dihydric phenols is their transformation to coloured products which stain polyolefins. Intense discolouration is particularly caused by derivatives of pyrocatechol which form red-coloured 1,2-benzo-
quinones. 1,4-Benzoquinones formed from alkylated hydroquinones are yellow or
orange. Both types of dihydric phenols rise by oxidation coupling and further consecutive reactions to brown discoloured compounds of the type of humic and
hymatomelanic acids[184].

Dihydric phenols interrupt an autooxidation chain by the reaction with $RO_2^{.}$
according to

$$InH + RO_2^{.} \longrightarrow In^{.} + ROOH$$
$$\lfloor RO_2^{.} \longrightarrow BQ + ROOH \qquad (10)$$

(BQ = benzoquinone)

The strong antioxidation efficiency of dihydric phenols and, at the same time, their
facile oxidation are connected with the presence of two hydroxyl groups in positions
1,2 or 1,4 of benzene nucleus. Generally, hydroquinones have lower values of redox
potential (500—660 mV[185]) than pyrocatechols (765—773 mV[186]). The values
decrease by the gradual alkylation of aromatic nucleus in both isomeric series and
may be well correlated with the steric and electronic factors of substituents. However, the redox potential value is not a sufficient condition to achieve the stabiliza-

tion ability: even the redox potentials of some pyrocatechol and hydroquinone derivatives, which exhibit only a very weak antioxidation efficiency[180, 181], are within the relatively broad limits of redox potentials (400–900 mV) given in the literature[4] as one of the necessary conditions for chain-breaking antioxidants to attain the antioxidation property.

Derivatives of hydroquinone are much weaker antioxidants than pyrocatechols[180–182]. These differences are connected with the mentioned difference of redox potentials, with the participation of hydroquinone derivatives in a side initiation reaction with ROOH[187, 188], and with differences in properties of benzoquinones formed.

A. Radical Intermediates and Quinoid Compounds

The product and mechanistic studies of oxidation of important pyrocatechols and hydroquinones by oxygen in an alkaline medium[189–195] give information on the types of radical intermediates formed. It was found by means of flow ESR methods[196, 197] that the radical CXXXVII arises in aquenous alkaline medium in the oxidation of 2-tert-butylhydroquinone CXXXVI, which is one of the most efficient hydroquinone antioxidants. As a secondary radical, the dianion radical CXLI was detected in a static arrangement, which is formed via intermediates

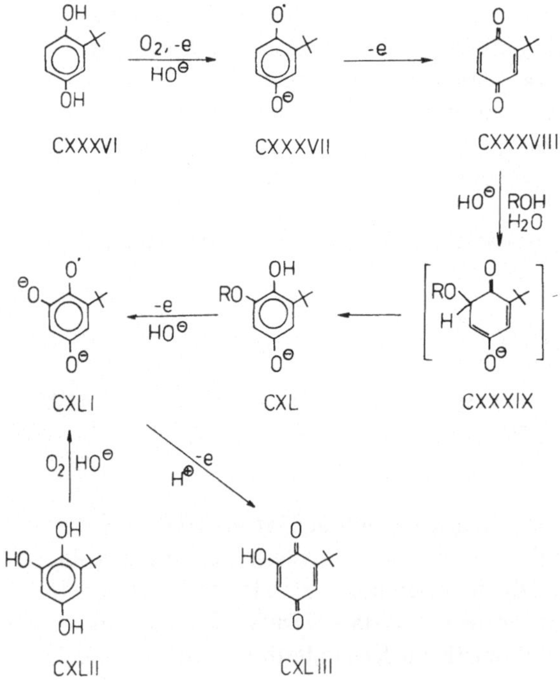

Scheme 17

CXXXIX and CXL (R = H). The reaction proceeds analogously in aqueous-alcoholic medium (in intermediates CXXXIX and CXL, R = Me or Et). Analysis of ESR spectrum of the secondary radical as well as the model oxidation of 2-tert-butyl-6-hydroxy-hydroquinone CXLII confirmed the structure of the dianion radical CXLI. Also the oxidation of other 2-tert-alkyl-, 2-sec.alkyl-, or 2-n-alkylhydroquinone proceeds by the same mechanism[198]. The oxidation can be carried out preparatively [194, 195, 198]; 2-hydroxy-6-alkyl-1,4-benzoquinones of type CXLIII are formed in the yield of 65 to 87%.

The course of oxidation of 4-tert-butylpyrocatechol CXLIV is shown in Scheme 18[196, 197]. In aqueous medium, the ion radical CXLV and the dianion radical CL which is derived from 2-hydroxy-5-tert-butyl-1,4-benzoquinone CLI are formed. The intermediates are CXLVII and CXLVIII (R = H). The course of oxidation is more complex in an aqueous-alcoholic medium. Formation of 4-tert-butyl-5-methoxy-1,2-benzosemiquinone CIL (R = Me) was found in this case. The oxidation of other 4-tert-alkylpyrocatechols also has an analogous course. In the preparative arrangement of oxidation, 4-tert-alkyl-1,2-benzoquinone CXLVI was not obtained, however, 2-hydroxy-5-tert-alkyl-1,4-benzoquinones of type CLI can be isolated in high yield[192–195].

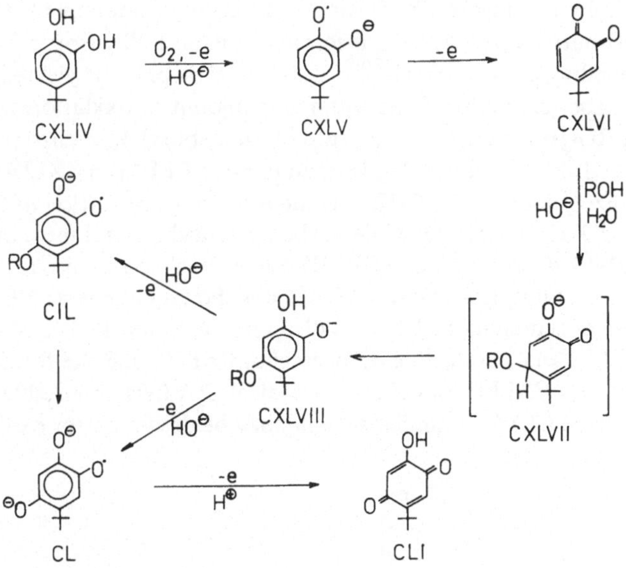

Scheme 18

It follows from Schemes 17 and 18 that different isomeric hydroxybenzoquinones are formed from mono-tert-alkylated pyrocatechols and hydroquinones. The position of newly introduced hydroxyl group is in accordance with data[199] dealing with the maximum spin density in isomeric semiquinones. The formation of different hydroxybenzoquinones is not probably the cause of differences in the antioxidation efficiency of derivatives of pyrocatechol and hydroquinone.

In contrast to monohydric phenols, also non-alkylated pyrocatechol or hydroquinone and their monomethyl derivatives are antioxidation effective. During
the oxidation in water-alcoholic alkaline medium, 2,5-dihydroxy-1,4-benzoquinone
CLII[189, 190, 193] and 2-hydroxy-5-methyl-1,4-benzoquinone[191] are formed from
pyrocatechol and 4-methylpyrocatechol, respectively. The oxidation of 2-methyl-
hydroquinone is more complex and more products are formed. Besides ion radicals·
CXXXVII and CXLI, also the ion radical CLIII was identified[198] in the study of
reaction mechanism. Intermediate CLIII corresponds to the formation of dimeric
hydroxybenzoquinone CLIV.

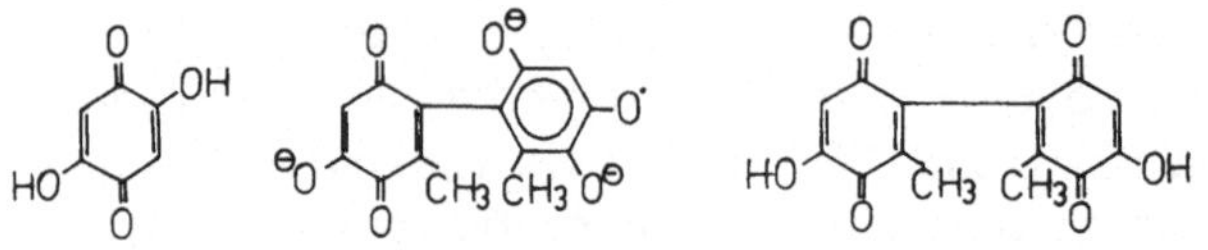

CLII   CLIII   CLIV

The oxidation of pyrocatechol and hydroquinone antioxidants in alkaline
medium does not exactly simulate the process which takes place during inhibited
oxidation. However, it allowed to identify radical intermediates and hydroxy-
quinonic compounds. It enabled the identification of products formed from 4-tert-
butylpyrocatechol CXLIV or 2-tert-butylhydroquinone CXXXVI by oxidation with
tert-butylhydroperoxide or tetrahydronaphthylhydroperoxide and with the corresponding peroxyradicals[200]. Both dihydric phenols yielded the corresponding
benzoquinones when the equimolar amount of oxidation agent was used. The excess
of ROOH gave rise to a mixture of 4-tert-butyl-1,2-benzoquinone CXLVI with
2-hydroxy-5-tert-butyl-1,4-benzoquinone CLI from CXLIV. Only 2-tert-butyl-1,4-
benzoquinone CXXXVIII was the product of oxidation of 2-tert-butylhydroquinone
CXXXVI in benzene, while in the tert-butyl alcohol medium also 2-hydroxy-6-tert-
butyl-1,4-benzoquinone CXLIII was formed as a by-product. The course of oxidation
of 4-tert-butylpyrocatechol CXLIV is shown in Scheme 19. It comprises the formation of semiquinone CLV, its disproportionation to 4-tert-butyl-1,2-benzoquinone
CXLVI and the starting pyrocatechol CXLIV, and the formation of hydroxybenzoquinone CXLI from mesomeric radical CLVI via an instable alkylperoxycyclohexadienone CLVII. A similar scheme may be drawn up for oxidation of 2-tert-butyl-
hydroquinone.

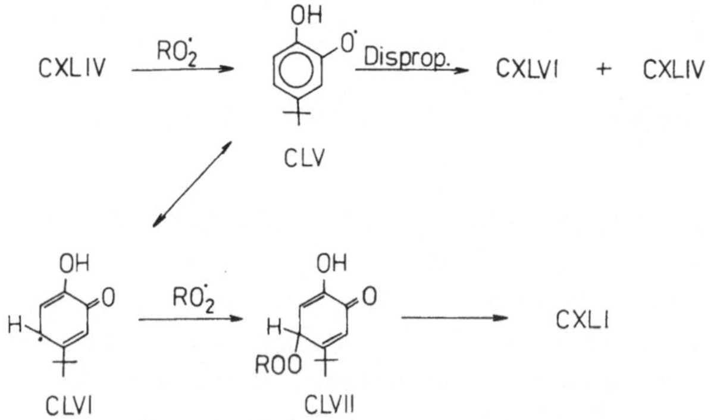

Scheme 19

Benzoquinones and hydroxybenzoquinones are formed from tert-butylated hydroquinone and pyrocatechol also in the course of inhibited oxidation of atactic polypropylene[200]. But these quinoic compounds are formed also in the direct oxidation by oxygen in trichlorobenzene at temperatures above 100 °C[200, 201]. This oxidation was negligible below 100 °C; it was observed, however, also at temperature over 110 °C in cumol[203, 204]. In this way, the formation of products having the same structure was experimentally proved in oxidations by $RO_2^\cdot$, ROOH, and oxygen. The direct oxidation of antioxidants by oxygen is undesirable, but it is unavoidable due to the facile oxidation of dihydric phenols. Not only alkylated but even halogenated bivalent phenols undergo the easy oxidation[201].

Analysis of the oxidation products of chain-breaking antioxidants from the group of dihydric phenols reveal that quinoid compounds are involved in the mechanism of their action. Due to the formation of benzoquinones and hydroxy-benzoquinones, the dihydric phenols are staining antioxidants. This fact reduces the technical importance of this in other ways very interesting group of antioxidants.

Quinoid compounds are excellent acceptors of electrons and form electron donor-acceptor (EDA) complexes as a consequence of low-lying unoccupied electronic energy levels[205]. The EDA complexes may be easily formed in interactions with phenolic or amine components of a stabilizing mixture, with other additives which have reactive H atoms, with $RO_2^\cdot$ radicals, or with some metallic impurities in polymers via π-orbital interactions. Quinones efficiently participate in oxidation of polymers by virtue of these processes.

The mechanism of H transfer from a donor (DH_2) is ionic and has two stages

$$DH_2 + Q \longrightarrow DH^+ + QH^- \longrightarrow D + QH_2 \tag{11}$$

Consequently, quinones act as outstanding oxidation and dehydrogenation agents[206]. This capacity of quinones may take part in a polymeric substrate in rather various ways. It comprises both dehydrogenation of polymer and the oxidation of present reactive compounds, particularly of antioxidants. For example, monohydric phenols, biphenyldiols, or dihydric phenols are very easily oxidized by quinones to the corresponding phenoxyls[207, 208], diphenoquinones or quinones[209–212], respectively. Also further products arise from binuclear phenolic antioxidants, e. g., spirocyclohexadienones of type XXIII from alkylidenebisphenols comprising a quaternary carbon atom in the bridge that connects phenol nuclei[40], or quinone methides, e. g. LVII from methylenebisphenols[75].

The effect of benzoquinones on thermal oxidation of hydrocarbons was investigated in detail on a liquid substrate. It was found[182] that 4-tert-alkyl-1,2-benzoquinones are at 60 °C more powerful retarders in the initiated oxidation of tetra-hydronaphthalene than 2-tert-alkyl-1,4-benzoquinones. The effect of both isomeric hydroxy-tert-alkylbenzoquinones of types CXLIII and CLI on this oxidation is entirely negligible. Unsubstituted 1,4-benzoquinone retards the oxidation of ethylbenzene[213]. However, the retardation effect of 2-tert-butyl- and 2,6-di-tert-butyl-1,4-benzoquinones CXXXVIII and XXII and, in particular, of 3,5,3'-tetra-tert-butyl-4,4'-diphenoquinone XX is much weaker than that of derivatives of quinone methide, e. g., XXIX. Moreover, benzoquinones resume their retardation

capacity generally only to 75 °C[31, 32, 214]. However, in the oxidation of atactic
polypropylene[94], a weak stabilization effect of 2,6-di-tert-butyl-1,4-benzoquinone
was observed at 120 °C.

The retardation action of benzoquinone derivatives in the thermal oxidation of
hydrocarbons may be connected not only with their ability to bind radicals $RO_2^{\cdot}$ in
the EDA complex, but also with the possibility of reacting with radicals $R^{\cdot}$. It has to
be considered indeed that in the oxidation of hydrocarbons under a sufficient
pressure of oxygen, $R^{\cdot}$ is oxidized to $RO_2^{\cdot}$ in a fast reaction. In spite of this fact, the
reaction of benzoquinones with $R^{\cdot}$ should be considered in the elucidation of oxidation retarded by quinones[15]. This process may have an exceptional importance in
the stabilization of polymers exposed to oxidation at low pressure of oxygen. The
stabilization process may be then regarded as analogous to reactions which proceed
with benzoquinones in the inhibition of polymerization. A structural study dis-
closed[217] that the stabilization capacity of quinones in the polymerization of
styrene decrease generally in the series 1,2-benzoquinone > $(HO)_n$-1,4-benzoquinone
> 1,4-benzoquinone > diphenoquinone. An ionic pair is formed first by electron
transfer in the reaction of benzoquinones with polymeric $R^{\cdot}$ according to[218]; an
effective inhibition species is semiquinone anion. The course of process, during which
benzoquinone is incorporated into the polymer chain, is influenced by the extent of
charge separation in ion-pair partners and by the polarity of the medium. As described in[219], 4-alkoxyphenoxyls CLVIII or 4-alkoxy-4-alkyl derivatives of
2,5-cyclohexadienone CLIX may be formed in the interaction of various benzoquinones with $R^{\cdot}$; the transformation of 2,6-di-tert-butyl-1,4-benzoquinone XXII
may serve as an example (Scheme 20).

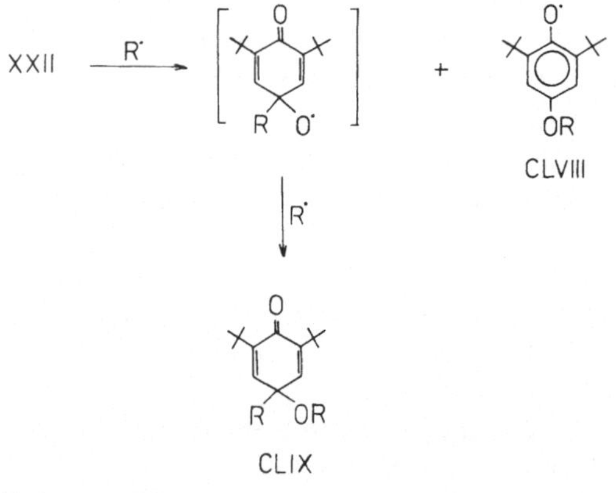

Scheme 20

It appears that the active participation of benzoquinone in the process of
thermal oxidation of hydrocarbons is essentially connected also with the formation of hydroquinoic structures. The reduction of 1,4-benzoquinone and anthra-
quinone in the course of thermal processing of polypropylene was proved by[220].
The major part of benzoquinone was transformed into hydroquinone; polypropylene
was the H-donor here. Due to this heat-induced reaction of benzoquinone, poly-

propylene was stable in the subsequent photooxidation. The similar course of thermal reduction of 3,6-di-tert-butyl-1,2-benzoquinone can be considered from investigations carried out with atactic[221] or isotactic polypropylene[222]. The fate of the polymeric radical arising by dehydrogenation of polypropylene with benzoquinones was not studied.

With respect to the presence of carbonyl group and the system of conjugated double bonds, participation of quinoid compounds in light-induced reactions, and, consequently, their active role in weathering of polymers have to be considered. Great attention was paid to the light-induced reactions of quinones generally[223] (see also former reviews[224–227]). A quinone Q forms, after absorption of electromagnetic energy, an excited singlet state and after an transition, a triplet state is formed which is probably responsible for subsequent chemical reactions[133, 228].

$$Q + h\nu \longrightarrow {}^1Q \longrightarrow {}^3Q \tag{12}$$

Quinones exhibit in the excited state a typical biradical behaviour[225–227]. They react intermolecularly with a substrate (i. e., with an H-donor) in the ground state. Cycloaddition of the excited quinoic system may be assumed in unsaturated polymers. In saturated polymers (RH), a free-radical elimination of H occur at bonds of low strength. Quinone ist reduced in this reaction to hydroquinone (QH_2) in the same way as in a thermally induced reaction.

$$^3Q + RH \longrightarrow {}^{\cdot}QH + R^{\cdot} \xrightarrow{RH} QH_2 + R^{\cdot} \tag{13}$$

The formed $R^{\cdot}$ rapidly reacts with oxygen giving $RO_2^{\cdot}$. It is possible that radicals $HO_2^{\cdot}$ will also be formed according to

$$^{\cdot}QH + O_2 \longrightarrow Q + HO_2^{\cdot} \tag{14}$$

It was proved experimentally that benzoquinones are readily photoreduced by phenolic or aromatic amine antioxidants. This process must be considered if mixtures of various chain-breaking antioxidants are applied in polymers. The photoreduction of 3,6-di-tert-butyl-1,2-benzoquinone by 2,4,6-tri-tert-butyl-phenol[229] or diphenylamine[230], was described. In both cases, semiquinone was formed from 1,2-benzoquinone and phenoxyl or amino radical arose from H-donors. It was even presented[231] that a stable complex may originate from 3,6-di-tert-butyl-1,2-benzoquinone and 3,6-di-tert-butylpyrocatechol, which is in equilibrium with 3,6-di-tert-butyl-2-hydroxyphenoxyl[232]. Photoreduction occurs also with bisnuclear quinones, e. g. diphenoquinone; its course is quantitatively different from that with mononuclear quinones due to the difference in redox potentials[223]. For example, the photoreduction of 3,5,3′,5′-tetra-tert-butyl-4,4′-diphenoquinone XX to 3,5,3′,5′-tetra-tert-butyl-4,4′-biphenyldiol XIX carried out in benzene was described[233]. Also biphenyl was isolated from the reaction mixture; this is the evidence that a non-activated aromatic nucleus may also act as the hydrogen donor.

It was assumed[133] that in the presence of molecular oxygen an energy transfer occurs from the excited triplet state of quinone under formation of singlet oxygen [reaction (9), $^3S^* =$ quinone]. Owing to the processes mentioned, quinones actively affect the photooxidation of polymers. Experimental data exist about the increased photodegradation of polyethylene[234], polyprophylene[235], cis-1,4-polyiso-prene[236, 237] or polystyrene[238, 239] in the presence of quinones. The process is obviously specifically influenced by the properties of polymer, the ability of quinones to quench triplet states of aromatic systems[240], and by photoinduced intramolecular processes substituted in benzoquinones. Consequently, benzo-quinones need not always accelerate the photooxidation. For example, the signi-ficant retardation effect of 2,6-di-tert-butyl-1,4-benzoquinone was found in the photooxidation of high-impact polystyrene[95].

Tert-butylated benzoquinone derived from common phenolic antioxidants undergo rather specific intramolecular reactions[241, 242]. The transformation is shown on the example of 2-tert-butyl-benzoquinone CXXXVIII in Scheme 21. Photolysis in the presence of an alcohol (which may be one of the products of polymer photooxidation) causes parallel photoreduction and intramolecular re-actions, in which a mixture of products is formed containing the corresponding tert-butylated hydroquinone CXXXVI, hydroquinone derivative CLXII with rearranged tert-butyl group, and a coumarone derivative CLXIII. Two semiquinones CLX and CLXI are formed during the photoreduction given in Scheme 21. The formation of CLXI is 5–7times faster[243]. The light-induced reactions of tert-butylated 1,4-benzoquinones occur also in the presence of aldehydes[244]; products are again the derivatives of hydroquinone and coumarone. The derivative of coumarone may arise also from 2,6-diphenyl-1,4-benzoquinone[245].

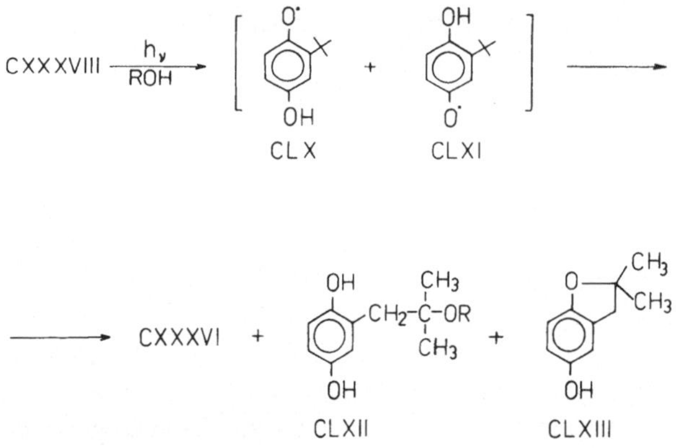

Scheme 21

Alkoxyalkylhydroquinones, e. g., CLXII, formed in the photoinduced transfor-mation of tert-butylated benzoquinones, exhibit strong antioxidation effects in polypropylene[214]. As long as compounds of this type are present in an oxidized polymer only in very low concentrations, they only retard the oxidation. This is probably the cause of the observed retardation effect of 2,6-di-tert-butyl-1,4-benzo-quinone in the oxidation of hihg-impact polystyrene[95].

Multinuclear products may be formed in phototransformations of unsubstituted benzoquinones or benzoquinones which do not contain substituents capable of intramolecular reactions. For example, dimers CLXIV or CXLV were formed from 2,5-dimethyl- or 2,6-dimethyl-1,4-benzoquinones[246]. Formation of other types of coupling products can be expectet, too.

CLXIV   CLXV

In connection with the discussion of the influence of quinoid compounds formed by transformations of antioxidants upon the long-term properties of polymers, carbon-black should be also mentioned. It is often used as a light-shield in the manufacturing of some articles from polyolefins. According to[247, 248], the structure of carbon-black is characterized by a polyconjugated system containing quinoid and phenolic functional groups, e. g., CLXVI. As a consequence of their presence, carbon-black may deactivate the radicals $RO_2^•$. Nevertheless, the antioxidation capacity of the sole carbon-black in polymers is very weak. However, its mixtures with chain-breaking antioxidants (above all with thiobisphenols or p-phenylenediamine derivatives) are very efficient protective systems for polyethylene and polypropylene[449, 250].

CLXVI

Carbon-black is present in polymers in a relatively large amount and, therefore, it may act even in an undesirable way. It can deactivate some antioxidants by accelerating their oxidation[90, 250-252], and influence in this way negatively the long-term properties of polyolefins. Loss of antioxidation efficiency of chain-breaking antioxidants may be also caused by their adsorption at the surface of carbon-black and at its acidity. According to[253], thiobisphenols are very resistent to these influences. However, the effect on polymer oxidation, extent of adsorption, and oxidation of antioxidants are very specifically dependent on the overall characteristics of carbon-black[251, 254]. Therefore, general conclusions cannot be drawn without experiments.

IV. Phenolic Sulphides

Polyolefins are very well protected against oxidation by phenolic sulphides[57, 255].
These compounds are more efficient than phenols containing a NH group in
the molecule as second reaction centre[255]. They possess moreover a technically outstanding property — they do not stain the polymer. Their stabilizing capacity is
closely connected with the mechanism of transformations during the antioxidation
action. Even when the oxidation transformation of phenolic sulphides and the
character of formed products have been studied only recently, in contradiction to
more simple monofunctional phenolic antioxidants, the obtained experimental data
already allow to make conclusions about intermediates and products of the transformations of important types of phenolic sulphides. The evidence about the antioxidation efficiency of some products of transformations and about their catalytic
effect in deactivation of hydroperoxides have been accumulated at the same time.

It is characteristic of phenolic sulphides thar their strong antioxidation effect
becomes evident above all at higher temperatures of oxidation. For example, they are
strong antioxidants in polypropylene at 180 °C[57, 255]; at this temperature, they
are more effective than alkylidenebisphenols or biphenyldiols. On the other hand,
their stabilization capacity studied at 65 °C in the oxidation of tetrahydronaphthalene is less marked[256, 257]. Significant features connected with the mechanism of
antioxidation action in polypropylene were already observed in the structural
studies: the impressive effect of the increased concentration of antioxidant on its
efficiency and the efficiency enhancement by the relative increase of sulphur content
in a molecule with respect to the content of OH. This fact was shown by comparison
of 4,4'-monothio-, 4,4'-dithio-, and 4,4'-trithiobisphenols, the relative efficiency of
which regularly increased in the given series by the factor of 1.4 for each sulphur
atom in a molecule[57, 255]. Another characteristic feature was observed: the oxidation of polypropylene stabilized with phenolic sulphides is strongly retarded after the
induction period. This phenomenon was not observed in polypropylene stabilized
with phenolic antioxidants not containing the sulphidic function. The behaviour in
the stages of oxidation of polypropylene after the induction period emphasizes the
importance of accumulating specific transformation products of antioxidants. Proofs
of the participation of thiobisphenol transformation products were found also in the
low-temperature oxidation of tetrahydronaphthalene: the stoichiometric factor as
well as the retardation capacity increased with the increasing content of sulphur in
the molecule[256, 257]. However, differences in the mechanism of antioxidant action
may be expected at ambient temperatures in comparison with a high-temperature
oxidation. Thus, for example, thiobisphenols completely lose their antioxidation
efficiency in tetrahydronaphthalene at 65 °C after gradual etherification, while an
impressive retardation action is still retained in polypropylene at 180 °C[258].

To elucidate the mechanism of transformation of phenolic sulphides, some data
can be used dealing with transformations of aliphatic sulphides, applied as synergistic additives, or with transformations of aromatic sulphides. However, the relationships should be extrapolated only very carefully. Further experimental data are
necessary to establish the whole transformation mechanism of phenolic sulphides.

Antioxidants with the structure of thiobisphenols[22, 23, 57, 256, 259—261] and hydroxybenzylsulphides[84, 255, 257, 262)] have been investigated more systematically. These types of antioxidant have two reaction centres of different reactivity in one molecule. The reactions considered for the phenolic part of molecule are characteristic above all by the interaction with radicals $RO_2^•$; evidence for the ability of the sulphidic part of molecule to decompose hydroperoxides is increasing. Consequently, phenolic sulphides can act as chain-breaking antioxidants and also as deactivators of hydroperoxides[5)]. In agreement with the previous text, the phenolic part of molecule is indicated as InH (15), the sulphidic part as D (decomposer of ROOH). The both mentioned reaction centres undergo a specific chemical transformation during the inhibition process via reactions (4) or (16) and (17).

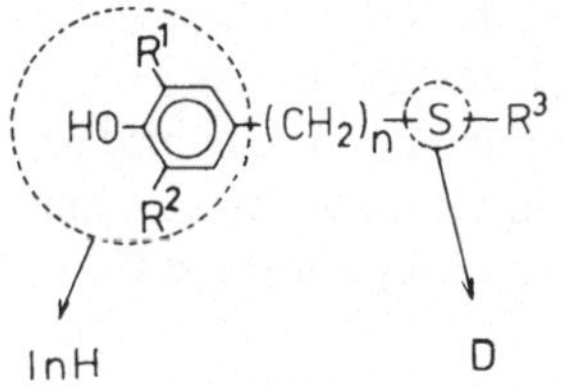

(15)

A. Phenoxy Radicals

Phenolic sulphides act as antioxidants in the oxidation of hydrocarbons at ambient temperature[256, 257)]. In this case, they inhibit the process chiefly by termination of oxidation chains according to (4) and phenoxy radicals In$^•$ are formed. Radicals of this type were obtained by independent reactions, e. g. by the one-electron electrochemical oxidation of 4,4'-thiobis(2,6-di-tert-butylphenol)[263)] or by oxidation of 4,4'-thiobis(3-methyl-6-tert-butylphenol) with benzoyl peroxide[264)]. Better understanding of the properties of phenoxy radicals derived from antioxidative important types of thiobisphenols was enabled by systematic studies[22, 23, 260)]. Phenoxyls were generated by cerium (IV) ammonium nitrate and lead (IV) oxide[260)], or under conditions simulating the autooxidation process by means of alkylperoxy radicals $RO_2^•$[22, 23)].

CLXVII CLXVIII

a: R^1 = Me, R^2 = tert-Bu

b: R^1 = R^2 = tert-Bu

Sulphur interrupts the conjugation between phenolic nuclei. Therefore, unpaired electron is delocalized mostly over one phenolic nucleus in phenoxyls formed from 2,2'-thiobisphenols CLXVII[23)]. Only about 15% of the density of unpaired electrons is delocalized to the sulphur of the sulphidic bridge. As a consequence, the stability of formed phenoxyls of the type CLXIX (Scheme 22) is lower in comparison with phenoxyls generated from 2,2'-biphenyldiols, where the structure allows delocalization of unpaired electron over both aromatic nuclei. The presence of the second sulphur atom in the bridge in phenoxyl derived from 2,2'-dithiobis(4-methyl-6-tert-butylphenol) has not already affected the delocalization of unpaired electrons in comparison with CLXIX.

Phenoxyls CLXXVI (Scheme 23) are formed from 4,4'-thiobisphenols of the type CLXVIII by oxidation with Ce(IV) and Pb(IV) compounds[260] and by the free or complex bonded radicals $RO_2^{.}$ [22]. The unpaired electron is always delocalized prevailingly over one aromatic nucleus[22, 260]; however, even partial delocalization to sulphur contributes to the stabilization of phenoxyl as follows from the comparison of splitting constants of meta protons of radicals derived from 4,4'-thiobisphenols CLXVIII and those in phenoxyls from 4,4'-methylenebis-phenols[22]. It was found that the presence of the second sulphur atom in the bridge of 4,4'-dithiobis(2,6-di-tert-butylphenol) causes the increase in values of the splitting constants in the phenoxyl formed. This is a difference from phenoxyls formed from 2,2'-thiobisphenols.

Phenoxyls are generated by the complex-bonded radicals $RO_2^{.}$[Co(III)] also from bis(3,5-di-tert-butyl-4-hydroxybenzyl)-sulphide CXCVIIa[22]. The bridge CH_2-S-CH_2 completely interrupts the mesomerism between both aromatic nuclei in the phenoxyl formed. However, although the atom S is separated from aromatic nuclei by a methylene group, its stabilization effect on the phenoxyls formed still occurs. This followed from the comparison with ESR spectra of phenoxyls from 4,4'-methylenebisphenols.

Free radicals of In$^{.}$ type are also formed in the reaction of some phenolic sulphides with tert-butylhydroperoxide(tert-BuOOH)[22, 23]. It is probable that an electron from the non-bonding pair of S atom is transferred to the peroxide bond, this splits and the liberated tert-butoxy radical reacts with thiobisphenol under formation of phenoxyl. The conception about the mechanism of formation of phenoxy radicals was experimentally confirmed: ESR signals of phenoxyls were not detected when the phenolic nuclei were bonded by a sulphoxide or sulphone bridge, instead of the sulphide bridge, thus excluding the transfer of electrons to peroxide. The reaction with tert-BuOOH proceeds very easily with 4,4'-thiobisphenols CLXVIII (R^1, R^2 = Me, Me; Me, tert-Bu; tert-Bu, tert-Bu), 4,4'-dithio-bis(2-methyl-6-tert-butyl-phenol), 4,4'-trithiobis(2-methyl-6-tert-butylphenol), and bis(3,5-di-tert-butyl-4-hydroxybenzylsulphide) CXCVIIa[22]. In the series of 2,2'-thiobisphenols CLXVII, the formation of phenoxyls was observed only with 2,2'-thiobis(4-methyl-6-tert-butylphenol)[23].

The above facts indicate a facile formation of phenoxyls from phenolic sulphides during the inhibition process as well as stabilization of phenoxyls due to the presence of sulphur atom in bridge even when sulphur is separated from the aromatic nucleus by a methylene group.

Phenolic sulphoxides and sulphones are formed in the transformation process of thiobisphenolic antioxidants (Chap. IV B). Sulphoxide restores the antioxidant capacity though to a lesser extent than that of sulphide[266, 267] while sulphone has no antioxidant properties. However, corresponding phenoxyls are formed from both these compounds[22, 260]; they are less stable than phenoxyls from the starting sulphide[260]. The group ArS(O)- stabilizes phenoxyl less effectively than the group ArS-[265].

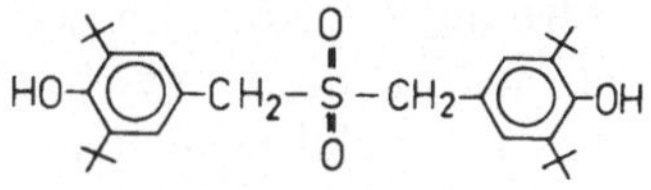

CLXXXIX

The stabilizing effect of the sulphonyl group in the phenoxyl derived from sulphone CLXXXIX is very remarkable. It is probably due to the stronger electronegative action of the $CH_2-SO_2-CH_2$ bridge in comparison with CH_2-S-CH_2[23]. It should be mentioned that the relatively high antioxidation efficiency was observed just with the above sulphone CLXXXIX[266].

Resonance effect connected with the formation of cyclohexadienonyl radical and its subsequent transformation into alkylperoxycyclohexadienone were proved in phenoxyls derived from monohydric phenols (Chap. II C.1). The possibility of participation of such reactions in transformations of phenolic sulphides of types CLXVII and CLXVIII can be assumed from data obtained in the oxidation by excessive alkylperoxyls[22, 23]. The recombination of alkylperoxyls occurs and active

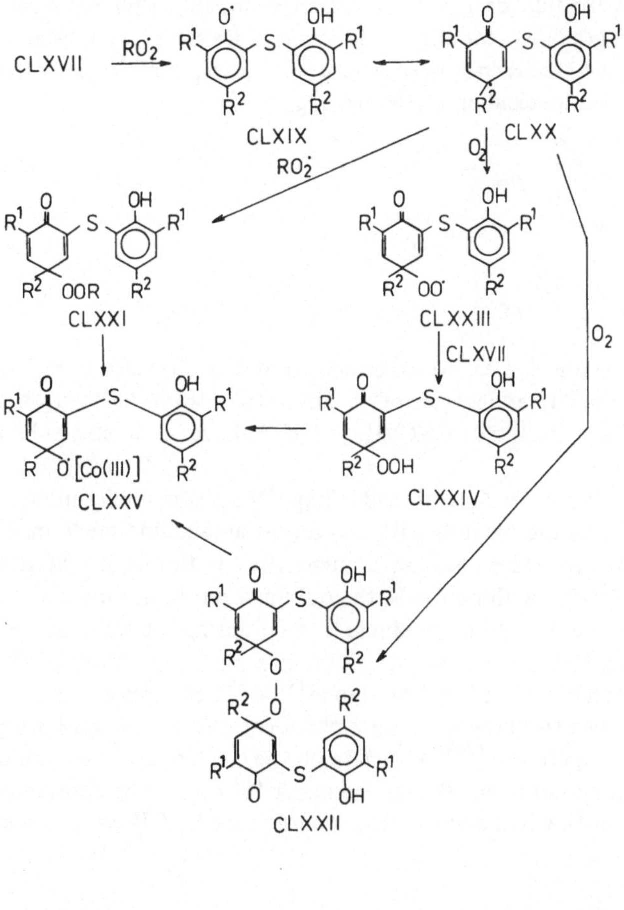

Scheme 22

oxygen is evolved. Free radicals characterized by octet signals in the ESR spectrum were formed from 2,2'-thiobisphenols CLXVII[23]. The radicals were interpreted as stabilized complexes of 2-(2-hydroxy-3,5-dialkylphenylthio)-4,6-dialkyl-2,5-cyclo-hexadienonyl-4-oxyls with Co(III) (CLXXV). Their formation under the given experimental conditions may be mapped by Scheme 22, which comprises the phenoxyl CLXIX and cyclohexadienonyl CLXX radicals, the reaction of the latter with $RO_2^{\cdot}$ giving alkylperoxycyclohexadienone CLXXI or, alternatively, with oxygen liberated in recombination of two $RO_2^{\cdot}$ giving either 4,4'-dioxybis(2,5-cyclohexadienone) CLXXII or cyclohexadienonylperoxy radical CLXXIII, transformation of the latter to cyclohexadienonylhydroperoxide CLXXIV by the reaction with thiobisphenol CLXVII, and the formation of CLXXV by catalytic cleavage of CLXXI, CLXXII or CLXXIV. The possibility of the catalytic cleavage of alkylperoxycyclohexadienone to cyclohexadienonyl radical was shown[268] using simpler model compounds. The obtained experimental data reveal that the formation of alkylperoxycyclohexa-dienones of type CLXXI from 2,2'-thiobisphenols CLXVII has to be considered during the inhibition process in polyolefins.

Radicals analogous to CLXXV were not obtained from 4,4'-thiobisphenols CLXVIII. This is the consequence of unfavourable steric conditions in the position 4,4' to hydroxy group which do not allow the reaction of mesomeric cyclohexa-dienonyl radicals with oxygen or $RO_2^{\cdot}$ [22]. However, the steric effect of dimethyl-enesulphidic bridge in CXCVIIa and the dimethylenesulphoxide and -sulphone bridges in its oxidation products enables such reactions and thus radicals were formed, ESR spectra of whose were interpreted[22] as Co(III) complexes of 2,5-cyclohexadienonyloxyls, e. g., CXC.

CXC

The radicals CXC were formed from CXCVIIa also by oxidation with oxygen bonded to Co(II)acetylacetonate[22]. This results from the electron density distribution in dibenzyl sulphide CXCVIIa, which behaves as an alkylsubstituted mononuclear phenol.

It has been mentioned (Chap. III A) that semiquinones are formed by oxidation of dihydric phenols with oxygen in an alkaline medium. Even when this oxida-tion procedure does not simulate the conditions of inhibited oxidation as true as the oxidation with peroxyls, the obtained results are indicative for the identification of structures of some products formed during the stabilization process. Thus, semi-quinones and hydroxysemiquinones are formed from alkyl derivatives of pyro-catechol[189] and hydroquinone[198, 269], corresponding to quinoid compounds formed from these dihydric phenols by action of tert-butylperoxyl or tert-butyl-hydroperoxide[200]. Also data on the oxidation of 4,4'-thiobisphenols CLXVIII carried out in an alkaline mediums[260] have to be considered from this view. Semi-quinone CXCI and hydroxysemiquinone CXCII were formed in the oxidation with oxygen. Oxidation of the same thiobisphenols by potassium hexacyanoferrate(III) in an aqueous methanolic sodium hydroxide solution gave either diphenosemiquinone CXCIII or the anion radical CXCIV, depending on the reaction conditions.

Phenoxyls represent the beginning of one route of thiobisphenol transformations during the inhibition process. Their further transformation is generally described in Chap. II A. It should be expected during the antioxidation process. Because they may be formed not only from the original phenolic sulphide but also from its various oxidation products, the number of possible products formed in the antioxidant-substrate system increases. It was also supposed[270] that the formation of phenoxyls is the cause of photo-yellowing, which was observed in weathered polyethylene stabilized by 4,4'-thiobis(3-methyl-6-tert-butylphenol) CLXVIIIa. The process of yellowing was inhibited by addition of a stabilizer from the group of derivatives of tetramethylpiperidine.

The formation and reactivity of radicals formed from thiobisphenols can be the cause of further intermolecular processes which may be used in the modification of polymers. For example, it was shown in[271] that di- and trithiobisphenols transfer chains in the polymerization of styrene and diene monomers. In this process, 2,2'-dithiobis(4-methyl-6-tert-butylphenol) was successfully bonded into the polymer chain.

B. Transformations on Sulphidic Bridge

Important information about the transformation of thiobisphenols under the conditions of inhibited oxidation were obtained on the basis of product analysis of model reactions. The detailed investigation was carried out with 4,4'-thiobis(2-methyl-6-tert-butylphenol) (CLXVIIIa), which is among the most effective thiobisphenolic antioxidants and tert-BuOOH was chosen as a simple model of polypropylene hydroperoxide. The reaction was followed at 20 °C in benzene solution and in the presence of the catalytic amount of Co(II)acetylacetonate[272]. Under these conditions, the phenolic antioxidant was attacked by both alkylperoxyls and hydroperoxide, so that its transformations were studied under conditions simulating the real system. The presence of more than 20 compounds was proved in the reaction mixture by means of TLC in the preparative arrangement of reaction. A part of the compounds formed was present in amounts which did not allow their isolation for identification purposes. The compounds present in larger amounts were isolated by the combination of repeated column chromatography and crystallization. The analysis of structures of isolated compounds revealed that their major part was formed in consequence of transformations at sulphur, while the phenolic function was retained. The formation of some of them was accompanied by elimination of sulphur atom from the molecule. The structures of main products and the pathways of their formation are shown in Scheme 23. The homolysis of thiobisphenol CLXVIIIa yields thiyl radicals CLXXVIIa and phenyl radicals CLXXVIIIa. Their recombination gives rise to a very strong antioxidant 4,4'-dithiobis(2-methyl-6-tert-butylphenol) (CLXXIXa) and to the sulphurless product, 3,3'-dimethyl-5,5'-di-tert-butyl-4,4'-biphenyldiol (XIXa) which is another effective phenolic antioxidant. Both products CLXXIXa and XIXa undergo further transformations. Biphenyldiol XIXa is oxidized by radicals $RO_2^{\cdot}$ to 3,3'-dimethyl-5,5'-di-tert-butyl-4,4'-diphenoquinone (XXa). According to[251, 273, 274], the mixture of compounds CLXXIXa und XXa is considered to be synergistic.

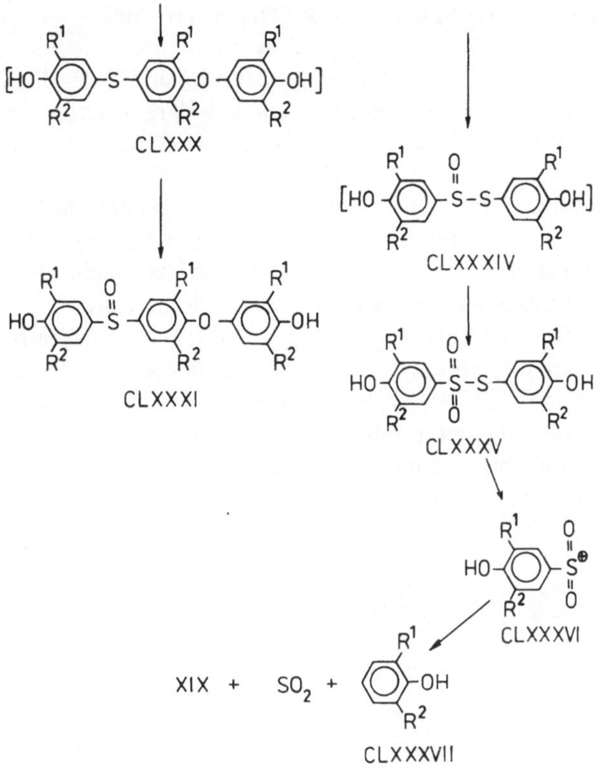

[HO—⬡—S—⬡—O—⬡—OH]

CLXXX

$$[HO—⬡—\overset{O}{\underset{||}{S}}—S—⬡—OH]$$

CLXXXIV

$$HO—⬡—\overset{O}{\underset{||}{S}}—⬡—O—⬡—OH$$

CLXXXI

$$HO—⬡—\overset{O}{\underset{\underset{O}{||}}{S}}—S—⬡—OH$$

CLXXXV

$$HO—⬡—\overset{O}{\underset{\underset{O}{||}}{S^{\oplus}}}$$

CLXXXVI

XIX + SO₂ + ⬡—OH

CLXXXVII

a : R^1 = Me, R^2 = t-Bu
b : R^1 = R^2 = t-Bu

Scheme 23

The formation of further compound CLXXXIa may be explained by recombination of phenyl CLXXVIIIa with phenoxyl CLXXVIa[272], which is primarily formed from phenol CLXCIIIa by the reaction with $RO_2^{\cdot}$ or ROOH. The probable intermediate was phenol CLXXXa and the final isolated product was 4-[(3-methyl-5-tert-butyl-4-hydroxyphenoxy)-3-methyl-5-tert-butylphenylsulphinyl]-2-methyl-6-tert-butylphenol CLXXXIa.

Organic sulphides are able to reduce hydroperoxides according to the general scheme[275]

$$R'SR' + ROOH \longrightarrow R'SOR' + ROH \tag{16}$$

Sulphoxides are formed in this reaction. Similarly thiolsulphinates arise from disulphides. Sulphoxide may originate also in a two-electron reduction by tert-butylperoxyl according to[275-277] (17)

$$R'SR' + \text{tert-BuO}_2^{\cdot} \longrightarrow R'SOR' + \text{tert-BuO}^{\cdot} \tag{17}$$

Some products of interaction of tert-BuOOH with thiobisphenol CLXVIIIa were probably formed according to these general reactions. From the reaction mixture, these were isolated[272] 4,4'-sulphinylbis(2-methyl-6-tert-butylphenol) CLXXXIIa and 4,4'-sulphonylbis(2-methyl-6-tert-butylphenol) CLXXXIIIa. 4-[(3-Methyl-5-tert-butyl-4-hydroxyphenylthio)-sulphonyl]-2-methyl-6-tert-butylphenol CLXXXVa is derived from dithiobisphenol CLXXIXa which belongs to primary products of trans-formation. The thiosulphinate CLXXXIVa which is less stable to thermal decomposi-tion and oxidation is assumed as an intermediate.

Further isolated products were 4-[(3-methyl-5-tert-butyl-4-hydroxyphenoxy)-sulphonyl]-2-methyl-6-tert-butylphenol CXCV and a dark brown crystalline com-pound CXCVI which was not identified. This compound is interesting because it was the only isolated product which contained sulphur, hydroxyl and carbonyl groups (presumably of a quinoid character) in the molecule at the same time.

CXCV

Separation of the reaction mixture after oxidation of CLXVIIIa was very diffi-cult. Semiquantitative data about the yields of individual transformation prod-ucts of thiobisphenols CLXVIIIa may give an idea about the ratios of individual re-actions which proceed simultaneously in one reaction mixture at ambient tempera-ture[272]:

Compound	XIXa	XXa	CLXXIXa	CLXXXIa	CLXXXIIa
Yield, %	3.5	1.2	2.3	4.7	58.2
Compound	CLXXXIIIa		CLXXXVa	CXCV	CXCVI
Yield, %	0.6		1.1	~0.5	~0.5

Sulphoxide CLXXXIIa was the main isolated product and the compounds containing sulphur in the molecule generally preponderated. DTA revealed that sulphoxides CLXXXIa (decomp. at 205 °C) and CLXXXIIa (decomp. at 180 °C) were the less thermally stable among these products.

Although the model study[272] was carried out at laboratory temperature, the reaction mixture contained in amounts which enabled isolation only the oxidation or thermally most stable compounds. The absence of some further possible oxidation products may be explained by their low stability. The thermal decomposition of non-phenolic derivatives of p-tolylsulphide proved[278] that the presence of a sulphinyl group weakens the S-S bond. The facility of thermolysis is characterized in this group of compounds by the following sequence (Ar = p-tolyl):

ArSSAr, ArSSO$_2$Ar, ArSO$_2$SO$_2$Ar $<$ ArSOSAr $<$ ArSOSO$_2$Ar $<$ ArSOSOAr.

Analogous relationships can be expected with phenolic sulphides. The unstable products of transformation are sources of thiyl, sulphinyl, and sulphonyl radicals, which may participate in the formation of sulphonate CXCV and other compounds present in the reaction mixture in very small amounts. It is possible that they are also the cause of formation of species which decompose catalytically tert-BuOOH (cf. also[259]).

Scott et al.[262] investigated the decomposition of cumylhydroperoxide by 4,4'-thiobis(2,6-di-tert-butylphenol) CLXVIIIb. They showed that CLXVIIIb reacts slowly at 75 °C and is transformed to the corresponding sulphoxide CLXXXIIb in a good yield at 95 °C.

Further data[259] were obtained in the study of tert-BuOOH decomposition by 4,4'-thiobis(2-methyl-6-tert-butylphenol) CLXVIIIa in chlorobenzene at 75 and 85 °C. At the tenfold molar excess of tert-BuOOH, the decomposition of tert-BuOOH exhibited two characteristic phases. The break on the kinetic curve, which indicated acceleration of tert-BuOOH decomposition, took place after about 2 mol of tert-BuOOH per mol CLXVIIIa had reacted. The decomposition had the similar characteristic course of reaction also at lower molar excess of tert-BuOOH. The total amount of decomposed tert-BuOOH considerably exceeded the stoichiometry of the process and the course of decomposition indicated that a specific action of an intermediate formed from the original antioxidant took part. It was found by liquid chromatography of the reaction mixture[259] that the gradual transformation of sulphide CLXVIIIa to sulphoxide CLXXXIIa occurred and that the maximum concentration of sulphoxide CLXXXIIa reached the value equimolar to original sulphide CLXVIIIa. Just after sulphide CLXVIIIa had disappeared, also sulphone CLXXXIIIa began to be formed and its maximum concentration in the mixture reached only roughly a half of that equivalent to original sulphide CLXVIIIa. However, another two compounds formed at the same time were detectable by TLC. It was independently shown that also sulphoxide CLXXXIIa decomposes a larger amount of tert-BuOOH than corresponds to the stoichiometry of sulphone CLXXXIIIa formation. Even in this case, two non-identified organic compounds, the same as in the reaction of sulphide CLXVIIIa, were the reaction products besides sulphone CLXXXIIIa. Sulphone CLXXXIIIa itself does not decompose tert-BuOOH under the given conditions. Accordingly, the transformation of sulphide CLXVIIIa (or sulphoxide CLXXXIIa) is to be considered a process leading to a species which catalytically decomposes tert-BuOOH. The structure of this species was not proved. An orientation evaluation of kinetics showed that the reaction rate of sulphide CLXIIIa with tert-BuOOH is approximately four times higher than that of sulphoxide CLXXXIIa.

Almost parallel to the product study[272], the reaction of cumyl hydroperoxide with 4,4'-monothio- to 4,4'-tetrathiobis(2,6-di-tert-butylphenols), sulphoxide CLXXXIIb, thiosulphonate CLXXXVb, and some other sulphurous compounds[261] was studied at 120 °C. It was stated from the kinetic and product analyses that the decomposition of cumyl hydroperoxide proceeded catalytically.

Considering that the formation of sulphoxides of type CLXXXII was independently observed under various conditions modelling the inhibition of polyolefin oxidation by 4,4'-thiobisphenols [259, 261, 262, 272] and, in addition, another phenolic compounds containing sulphur were identified[272], sulphoxides and their consecutive transformations should be regarded as key events in the mechanism of formation of species causing the catalytic decomposition of ROOH. These species are formed during the transformation process of the original thiobisphenol, similarly as it is with aliphatic sulphides[279, 280]. Consequently, the catalytic effect on ROOH decomposition, which might be superficially ascribed to the originally present thiobisphenol is the result of the complex process of transformation. The final effect depends on the amount and efficiency of catalytically active species. In studies[261, 262], sulphoxide CLXXXIIb derived from CLXVIIIb is stated as the only sulphur-containing product of transformation. According to[301, 302], catalytically active species are formed via sulphoxides in the series of aliphatic compounds. However, thiolsulphinates or thiolsulphonates derived from 4,4'-dithiobisphenols CLXXIX, e. g., compounds CLXXXIV or CLXXXV, may be considered as more probable precursors of catalytic species in the thiobisphenol series. This follows also from the comparison of antioxidation efficiences of 4,4'-thiobis(2-methyl-6-tert-butylphenol) and 4,4'-dithiobis)2-methyl-6-tert-butylphenol) in the stabilization of isotactic polypropylene[57, 267] where dithiobisphenol CLXXIXa exceeded the efficiency of monothiobisphenol CLXVIIIa by the factor 1.4.

Thermochemistry of the formed sulphoxides or thiolsulphinates may be regarded as decisive for the formation of catalytically active species from the originally applied sulphide[274]. However, the reaction temperature is very important. It was inferred[261] from the products of cumyl hydroperoxide decomposition at 120 °C that 4,4'-monothio- to 4,4'-tetrathiobis(2,6-di-tert-butylphenols) decompose ROOH by the same mechanism. The effect of 4,4'-dithiobis(2,6-di-tert-butylphenol) CLXXIXb was followed in more detail. The sulphurless compounds 3,5,3',5'-tetra-tert-butyl-4,4'-biphenyldiol XIXb and 2,6-di-tert-butylphenol CLXXXVIIb were found in a high yield as nonvolatile products. The described reaction mixture[261] was, however, much simpler than that described in [272]. In accordance with the thermal instability (at temperatures above 100 °C) of compounds expected as the intermediates of dithiobisphenol transformation by hydroperoxide, i. e. of sulphoxides, thiolsulphinates, and thiolsulphonates[278, 281−283], sulphur dioxide was proposed[261] as the catalyst of ROOH decomposition. Its formation was proved by means of gas-liquid chromatography (GLC) directly in the reaction mixture and by the model thermolysis of CLXXXVb at 120 °C. It seems, therefore, that thiolsulfonate CLXXXVb is the straight precursor of SO_2. The complex mechanism of transformation given in Scheme 23 may be extended also to 4,4'-trithiobis(2,6-di-tert-butylphenol) and 4,4'-tetra-thiobis(2,6-di-tert-butylphenol) provided polysulphides are oxidized primarily at the terminal sulphur atom[284]. The formation of mono- to trithiolsulphonates, which decompose according to[285] heterolytically to a sulphonyl cation CLXXXVI and liberate SO_2 in the subsequent step, may then be generally assumed in the course of transformations. Another possibility for formation of SO_2 is oxidation of thiyl radical CLXXVIIa to sulphonyl radical CLXXXVIIIa, even when the oxidation of S-centered radicals generally appears to be slower than that of carbon-centered radicals[286].

The product study[272] and the kinetic study[261] form the experimental basis for the discussion of complex transformations of thiobisphenolic antioxidants. Their retardation effect in polypropylene after the induction period and the mechanism of formation of intermediates or products catalytically active in ROOH decomposition indicate the reactivity of both different centres in molecule and provide partial data for the discussion of intramolecular synergism of thiobisphenols[5, 57, 258, 267].

The intramolecular contribution of two inhibition mechanism was indicated in the stabilization of polypropylene by bis(3,5-di-tert-butyl-4-hydroxybenzyl)sulphide and bis(3,5-di-tert-butyl-2-hydroxybenzyl)sulphide[255]. The possibility of catalytic decomposition of ROOH, as a parallel process to the deactivation of radicals $RO^{\cdot}_2$, was postulated in the kinetic investigation of autooxidation of pentaerythritol esters inhibited by 2-hydroxy-3-(N,N-dimethylaminomethyl)-5-methoxybenzyl benzyl sulphide[287]. The catalytic action of 2-hydroxybenzyl phenyl sulphide in the decomposition of cumyl hydroperoxide was indicated in another kinetic study[288]. The mechanism of action of benzyl sulphides with cumyl hydroperoxide was investigated in detail by Scott[262]. He used 3,5-di-tert-butyl-4-hydroxybenzyl alkyl sulphide CXCVIIb and bis(3,5-di-tert-butyl-4-hydroxybenzyl) sulphide CXVIIa and analyzed the specific features of this catalytic process. It was stated that the catalytically active species is an inorganic acidic compound, probably SO_3.

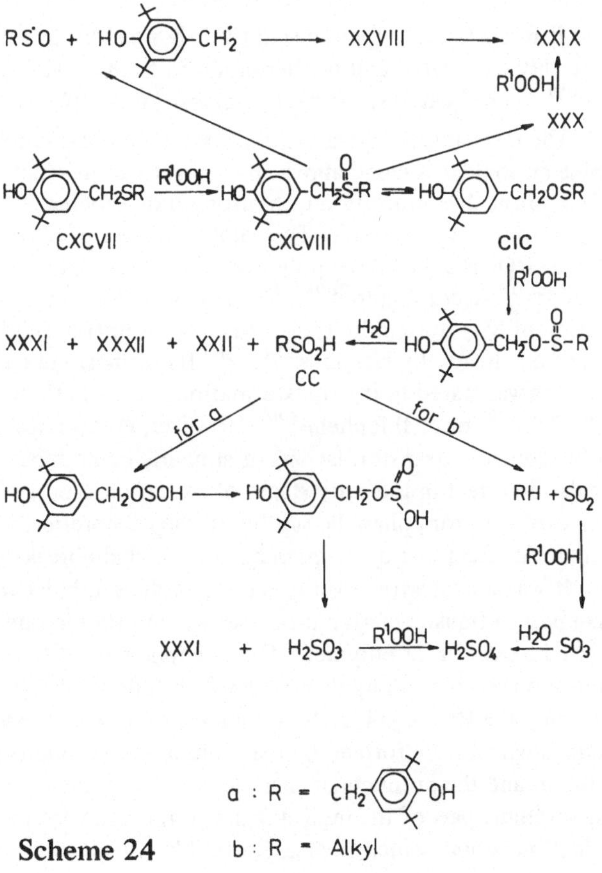

Scheme 24

An induction period was observed in the decomposition of cumyl hydroperoxide in chlorobenzene at 70 and 110 °C in the presence of phenolic sulphides CXCVIIa,b[262]. This was a substantial difference with respect to the behaviour of 4,4'-thio-bis(2,6-di-tert-butylphenol) CLXVIIIb which decomposed ROOH under the same conditions without induction period. The result indicates a mechanistic distinction in the action of both types of phenolic sulphides. In the mechanism of transformations of benzyl sulphide CXCVIIb, there are assumed (Scheme 24) the formation of sulphoxide CXCVIII and the intermediary formation of CIC followed by oxidation and formation of sulphinic acid CC. Further transformation of the acid CC depends on the character of R. If R = 3,5-di-tert-butyl-4-hydroxybenzyl, as it is in the formation of CC from CXCVIIa, the total elimination of the sulphurous part of molecule may occur and the transformation products of phenolic or quinoid character may be formed: 3,5-di-tert-butyl-4-hydroxybenzyl alcohol XXXI, the corresponding aldehyde XXXII, and 2,6-di-tert-butyl-1,4-benzoquinone XXII were identified. Another possible sulphurless product is 4,4'-ethylenebis(2,6-di-tert-butyl-phenol) XXVIII, which was isolated in small amounts in its oxidized form as 3,5,3',5'-tetra-tert-butyl-4,4'-stilbenequinone (XXIX). Quinone methide XXX formed by thermolysis of sulphoxide CXCVIII, may be also the precursor in formation of XXIX. According to[66], XXX is further oxidized by hydroperoxides to XXIX.

If R is alkyl in sulphinic acid CC (obtained by transformation of CXCVIIb), SO_2 is readily eliminated and further oxidized to SO_3. This compound is according to[289] the real catalyst of the decomposition of ROOH (Scheme 24).

The mechanisms of transformations of thiobisphenol and hydroxybenzyl sulphide antioxidants based above all on results of product studies justify their catalytic effect in the decomposition of hydroperoxides being explained by formation of sulphur dioxide or trioxide. The catalytic effect of SO_2 on the decomposition of various ROOH's, including polypropylene hydroperoxide, was experimentally confirmed[290]. According to[289], SO_2 possesses also a chain-breaking function. Formation of SO_2 during the transformation of phenol sulphide antioxidants was supposed already by Hawkins[274, 291]. Its formation as an antioxidatively active product was stated in the transformations of aliphatic sulphurous antioxidants[289, 290, 292-299] or of thiophene[300]. However, even radical intermediates may occur in the course of transformations of aliphatic compounds which exhibit their pro-oxidation effect in the early stages of oxidation process[295]. If such intermediates can arise also from phenolic sulphides, they theoretically need not exhibit a pro-oxidation effect due to the presence of the chain-breaking part of molecule.

It was shown with two types of phenolic sulphides that both reaction centres take part in transformation processes which proceed under the conditions of inhibited oxidation of polymers. The phenolic part of the molecule participates in the inhibition mechanism by processes which follow from its chain-breaking function. Not only the starting phenolic sulphides react in this way, but also the phenolic compounds which were formed by reactions at the second centre, i. e., at sulphidic sulphur, and the products of their consecutive transformations. Transformations of the sulphidic part of the molecule are connected with preventive mechanism of anti-oxidation action, which is characterized by the decomposition of hydroperoxides.

The overall reaction scheme of the transformations of phenolic sulphides is complex. However, it indicates the exceptional importance of knowledge of the main transformation products of the originally applied antioxidant during the inhibition process and of consequences of these transformations.

V. Conclusions

The functional transformation is an integral part of the stabilization capacity of antioxidants and has to be considered in the discussion of relationships between structure and efficiency. The analyses indicate that it is impossible to elucidate the complex mechanism of action of antioxidants without the knowledge of chemical and photochemical properties of products. The transformation products are cummulated in a polymer und play a specific role in various stages of its ageing. In this complex process, compounds with initiation, retardation, or antioxidant properties may be formed from original antioxidants. These compounds participate in the integral stabilization capacity of the additive used under real degradation conditions. This fact explains the differences in overall properties found under different conditions.

Monohydric phenols, the most applied group of phenolic antioxidants, are changed via phenoxyls into more types of products. The C–C coupling reactions leading to the antioxidation-efficient multinuclear phenols play a favourable role. The formation of quinone methionid compounds which stain polyolefins cannot be avoided in the transformation process. However, these compounds exhibit a retardation effect in thermal oxidation and are able to quench singlet oxygen. The least favourable properties have alkylperoxycyclohexadienones and dioxycyclohexadienones, which initiate both the thermal and photochemical oxidation. Products of their subsequent transformations are either inactive or have a weak retardation effect.

An important group involves quinoid compounds, most often benzoquinones and diphenoquinones. They originate from all typs of phenolic antioxidants either as the primary product, or by subsequent transformations, and may retard the oxidation. More complex relationships apply in light-induced processes.

It holds specifically for antioxidants with the structure of thiobisphenols and hydroxybenzyl sulphides that their active role in stabilization of polyolefins is not only accompanied by chemical transformations, but it is even determined by these transformations. Thus, phenolic sulphides, in addition to their chain-breaking function, are also precursors of active species which catalytically decompose hydroperoxides. They can serve as an example of compounds in which the transformation product of the original additive is an active antioxidant.

The knowledge of properties of transformation products of phenolic antioxidants is necessary also from the standpoint of hygiene and toxicology. Legislative approvals of antioxidants for alimentary and cosmetic purposes are granted for the present irrespective to the possible physiologic effects of their transformation products. At the same time, the product studies show the path of further development of phenolic antioxidants, in particular of multifunctional systems with the possibility of autosynergistic behaviour.

VI. Registered Trade Names and Formulae of Commercial Antioxidants Cited in the Text

Trade Name	Producer	Formula
Good Rite® 3114	B. F. Goodrich Chemical Co., Cleveland, Ohio, USA	1,3,5-Tris(3,5-ditert.butyl-4-hydroxybenzyl)cyanuric acid
Ionox® 330	Shell Chemical Co., New York, N. Y., USA	1,3,5-Trimethyl-2,4,6-tris(3,5-ditert.-butyl-4-hydroxybenzyl)-benzene
Irganox® 1010	Ciba-Geigy AG, Basle Switzerland	Methylene [3-(3,5-diterc.butyl-4-hydroxyphenyl)propionate] methane
Irganox® 1076	Ciba-Geigy AG, Basle Switzerland	Octadecyl 3-(3,5-ditert.butyl-4-hydroxyphenyl)propionate
Melilox®	Ciba-Geigy AG, Basle Switzerland	Methyl 3-(3,5-ditert.butyl-4-hydroxyphenyl)propionate

VII. References

1. Heller, H. J., Blattmann, H. R.: Pure Appl. Chem. *36*, 141 (1973)
2. Reich, L., Stivala, S. S.: Autoxidation of hydrocarbons and polyolefins. Kinetics and mechanism. New York: Dekker 1969
3. Scott, G.: Atmospheric oxidation and antioxidants. Amsterdam: Elsevier 1965
4. Pospíšil, J.: Antioxidanty. Praha: Academia 1968
5. Pospíšil, J.: Chain-breaking antioxidants in polymer stabilization. In: Developments in polymer stabilization. Scott, G. (ed.). London: Applied Science Publishers 1979, Volume 1, pp. 1–37
6. Pospíšil, J., Rotschová, J.: Rév. Gen. Caoutch. Plast. *54*, No 567, 72, No 568, 73, No 569, 131 (1977)
7. Pospíšil, J.: Pure Appl. Chem. *36*, 207 (1973)
8. Pospíšil, J.: Elucidation of structure-effectiveness relations of phenolic antioxidants in polypropylene. In: Kinetics and mechanism of polyreactions. IUPAC Intern. Symposium on Macromolecular Chemistry, Plenary and main lectures volume. Budapest: Akadémiai Kiadó 1971, pp. 789–808
9. Pospíšil, J.: Plaste Kaut. *24*, 396 (1977)
10. Pokhodenko, V. D., Khizhnyi, V. A., Bidzilya, V. A.: Usp. Khim. *37*, 998 (1968)
11. Forrester, A. R., Hay, J. M., Thomson, R. H.: Organic chemistry of stable free radicals. London-New York: Academic Press 1968
12. Altwicker, E. R.: Chem. Rev. *67*, 475 (1967)
13. Ershov, V. V., Nikiforov, G. A., Volodkin, A. A.: Prostranstvenno-zatrudnennye fenoly. Moscow: Khimiya 1972
14. Mc Gowman, J. C., Powell, T.: J. Chem. Soc. *1960*, 238

15. Waters, W. A.: Homolytic oxidation processes. In: Progress in organic chemistry, Volume 5. London: Butterworths 1961, pp. 1–45
16. Mahoney, L. R.: Angew. Chem. *81*, 555 (1969)
17. Rieker, A., Müller, E., Beckert, W.: Z. Naturforsch. *17b*, 718 (1962)
18. Rieker, A., Kessler, H.: Z. Naturforsch. *21b*, 939 (1966)
19. Griva, A. P., Denisov, E. T.: Intern. J. Chem. Kinet. *5*, 369 (1973)
20. Neiman, M. B., Mamedova, Yu. G., Blenke, P., Butchatchenko, A. L.: Dokl. Akad. Nauk SSSR *144*, 392 (1962)
21. Westfahl, J. C., Carman, C. J., Layer, R. W.: Rubber. Chem. Technol. *45*, 402 (1972)
22. Tkáč, A., Omelka, L., Jiráčková, L., Pospíšil, J.: Org. Magnet. Resonance in press
23. Tkáč, A., Omelka, L., Jiráčkova, L., Pospíšil, J.: Org. Magnet. Resonance in press
24. Müller, E., Ley, K., Kriedasch, W.: Chem. Ber. *87*, 1605 (1954)
25. Mahoney, L. R., Weiner, S. A.: J. Amer. Chem. Soc. *94*, 585, 1412 (1972)
26. Matsuura, T., Nishinaga, A., Cahnmann, H. J.: J. Org. Chem. *27*, 3620 (1962)
27. de Jonge, C. R. H. I., Hageman, H. J., Huysmans, W. G. B., Mijs, W. J.: Am. Chem. Soc., Org. Coatings Plastics Div. *34*, 107 (1974)
28. Rieker, A., Zeller, N., Schurr, K., Müller, E.: Justus Liebigs Ann. Chem. *697*, 1 (1966)
29. Müller, E., Mayer, R., Narr, B., Rieker, A., Scheffer, K.: Justus Liebigs Ann. Chem. *645*, 25 (1961)
30. Taimr, L., Pivcová, H., Pospíšil, J.: Chem. Ind. (London) *1975*, 747
31. Buben, I., Pospíšil, J.: Collect. Czech. Chem. Commun. *40*, 977 (1975)
32. Buben, I., Pospíšil, J.: Collect. Czech. Chem. Commun. *40*, 987 (1975)
33. Horswill, E. C., Ingold, K. U.: Can. J. Chem. *44*, 269 (1966)
34. Kharash, M. S., Yoshi, B. S.: J. Org. Chem. *22*, 1439 (1957)
35. Horswill, E. C., Ingold, K. U.: Can. J. Chem. *44*, 263 (1966)
36. Yohe, G. R., Hill, D. R., Dunbar, J. E., Scheidt, F. M.: J. Amer. Chem. Soc. *95*, 2688 (1953)
37. Waters, W. A., Wickham-Jones, C.: J. Chem. Soc. *1952*, 2420
38. Bourdon, J., Calvin, J.: J. Org. Chem. *22*, 101 (1957)
39. Chandross, E. A., Kreilick, R.: J. Amer, Chem. Soc. *85*, 2530 (1963); *86*, 117 (1964)
40. Becker, H. D.: J. Org. Chem. *34*, 1203 (1969)
41. Kurechi, T., Senda, H.: Eisei Kagaku *23*, 267 (1977); Chem. Abstr. *88*, 190281e (1978)
42. Waring, A. J.: Cyclohexadienones. In: Advances in alicyclic chemistry. Hart, H., Karabatsos, G. J. (eds). New York: Academic Press, Vol. 1, 1966, pp. 129–256
43. Waring, A. J.: Österr. Chem. Ztg. *68*, 232 (1967)
44. Ingold, K. U.: Can. J. Chem. *41*, 2816 (1963)
45. Bickel, A. F., Kooyman, E. C.: J. Chem. Soc. *1953*, 3211
46. Becconsall, J. K., Clough, S., Scott, G.: Trans. Faraday Soc. *56*, 459 (1960)
47. Huysmans, W. G. B., Waters, W. A.: J. Chem. Soc. B. *1966*, 1047
48. Bauer, R. H., Coppinger, G. M.: Tetrahedron, *19*, 1201 (1963)
49. Taimr, L., Pivcová, H., Pospíšil, J.: Collect, Czech. Chem. Commun. *37*, 1912 (1972)
50. Yoshikawa, Y., Kumanotani, J.: Makromol. Chem. *134*, 33 (1970)
51. Magnusson, K.: Acta Chem. Scand. *18*, 759 (1964)
52. Magnusson, K.: Acta Chem. Scand. *20*, 2211 (1966)
53. Sirimewan, K. W., Scott, G.: Europ. Polym. J. *14*, 835 (1978)
54. Hatchard, W. R., Lipscomb, S. F. W.: J. Amer. Chem. Soc. *80*, 3636 (1958)
55. Zikmund, L., Taimr, L., Čoupek, J., Pospíšil, J.: Europ. Polym. J. *8*, 83 (1972)
56. Pospíšil, J., Kotulak, L., Halaška, V.: Europ. Polym. J. *7*, 33 (1971)
57. Jiráčkova, L., Pospíšil, J.: Europ. Polym. J. *8*, 75 (1972)
58. Cook, C. D., Norcross, B. E.: J. Amer. Chem. Soc. *78*, 3797 (1956)
59. Becker, H. D.: J. Org. Chem. *30*, 982 (1965)
60. Bennett, J. E.: Nature *186*, 385 (1960)
61. Cook, C. D., Norcross, B. E.: J. Amer. Chem. Soc. *81*, 1176 (1959)
62. Brodskij, A. I., Pokhodenko, V. D., Khizhnyi, V. A., Kalibabkhuk, N. N.: Dokl. Akad. Nauk SSSR *169*, 339 (1966)
63. Stebbins, R., Sicilio, F.: Tetrahedron *26*, 291 (1970)

64. Kudinova, L. I., Volodkin, A. A., Ershov, V. V.: Izv. Akad. Nauk SSSR, otd. khim. *1978*, 1661
65. Cook, C. D., Nash, N. G., Flanagan, H. R.: J. Amer. Chem. Soc. *77*, 1783 (1955)
66. Filar, J. R., Weinstein, S.: Tetrahedron Lett. *1960*, 9
67. Coppinger, G. M.: J. Amer. Chem. Soc. *86*, 4385 (1964)
68. Loy, B. R.: J. Org. Chem. *31*, 2386 (1966)
69. Khudyakov, A. V., de Jonge, C. R. H. I., Levin, P. P., Kuzmin, V. A.: Izv. Akad. Nauk SSSR, otd. khim. *1978*, 1492
70. Müller, E., Mayer, R., Heilmann, V., Scheffer, K. : Justus Liebigs Ann. Chem. *645*, 66 (1961)
71. Müller, E., Schick, A., Mayer, R., Scheffer, K.: Chem. Ber. *93*, 2649 (1960)
72. Ayscough, P. B., Russell, K. E.: Can. J. Chem. *45*, 3019 (1967)
73. Kudinova, L. I., Volodkin, A. A., Ershov, V. V.: Izv. Akad. Nauk SSSR, otd. khim. *1978*, 2797
74. Samsonova, L. V., Taimr, L., Pospíšil, J.: Angew. Makromol. Chem. *65*, 197 (1977)
75. Brieskorn, C. H., Kallmayer, H.-J.: Justus Liebigs Ann. Chem. *741*, 124 (1970)
76. Bartlett, P. D., Rüchardt, D.: J. Amer. Chem. Soc. *82*, 1756 (1960)
77. Kharash, M. S., Joshi, B. S.: J. Org. Chem. *22*, 1435 (1957)
78. Brieskorn, C. H., Ullmann, K.: Chem. Ber. *100*, 618 (1967)
79. Steenlink, C., Fitzpatrick, J. D., Kispert, L. D., Hyde, J. S.: J. Amer. Chem. Soc. *90*, 4354 (1958)
80. Bartlett, P. D., Funahashi, F.: J. Amer. Chem. Soc. *84*, 2596 (1962)
81. Greene, F. D., Adam, W., Coutrill, J. E.: J. Amer. Chem. Soc. *83*, 3461 (1961)
82. Greene, F. D., Adam, W.: J. Org. Chem. *28*, 3550 (1963)
83. Dubinskij, V. Z., Beljakov, V. A., Roginskij, V. A., Miller, V. B.: Izv. Akad. Nauk SSSR, otd. khim. *1975*, 51
84. O'Shea, F. X.: Advan. Chem. Ser. *85*, 126 (1968)
85. Taimr, L., Pospíšil, J.: Angew. Makromol. Chem. *28*, 13 (1973)
86. Yang, N. C., Castro, A. J.: Amer. Chem. Soc. *82*, 6208 (1960)
87. Chandross, E. A.: J. Amer. Chem. Soc. *86*, 1263 (1964)
88. Koch, J.: Angew. Makromol. Chem. *20*, 7 (1971)
89. Koch, J.: Angew. Makromol. Chem. *20*, 21 (1971)
90. Dann, H., Gilbert, S. G., Giacin, J.: J. Amer. Oil Chemist's Soc. *51*, 404 (1974)
91. Popova, G. S.: Zhur. Prikl. Khim. *46*, 419 (1973)
92. Kaimai, T.: Sekiyu Gakkai Shi *21*, No 2, 89 (1978); Chem. Abstr. *89*, 149065 v (1978)
93. Wagner, H. V., Compper, R.: Quinone methides. In: The chemistry of the quinonoid compounds. Patai, S. (ed.). London: John Wiley & Sons 1974, pp. 1145–1178
94. Buben, I., Pospíšil, J.: J. Polym. Sci., Symposia *57*, 261 (1976)
95. Virt, J., Rosík, L., Kovářová, J., Pospíšil, J.: Eur. Polym. J., (submitted)
96. Taimr, L., Pospíšil, J.: Tetrahedron Lett. *1972*, 4279
97. Lerchová, J., Nikiforov, G. A., Pospíšil, J.: J. Polym. Sci., Symposia *57*, 249 (1976)
98. Volodkin, A. A., Ershov, V. V., Gorodetskaya, N. N., Tupikina, A., Kudinova, L. I.: Dokl. Akad. Nauk SSSR *227*, 896 (1976)
99. Schuster, D. I., Krull, I. S.: Molecular Photochem. *1*, 107 (1969)
100. Krull, I. S., Schuster, D. I.: Tetrahedron Lett. *1968*, 135
101. Becker, H. D.: J. Org. Chem. *32*, 2115 (1967)
102. Matsuura, T., Ogura, K.: Bull. Chem. Soc. Japan *42*, 2970 (1969)
103. Taimr, L., Pospíšil, J.: Angew. Makromol. Chem. *52*, 31 (1976)
104. Dubinskij, V. Z., Roginskij, V. A., Miller, V. B., Shlyapnikova, I. A.: Izv. Akad. Nauk SSSR, otd. khim. *1975*, 1180
105. Chien, J. C. W., Boss, C. R.: J. Polym. Sci. A–1,*5*, 1683 (1967)
106. Bickel, A. F., Kooymann, E. C.: J. Chem. Soc. *1957*, 2415
107. Moore, R. F., Waters, W. A.: J. Chem. Soc. *1954*, 246
108. Hey, M. E., Waters, W. A.: J. Chem. Soc. *1955*, 2753
109. Hey, M. E., Waters, W. A., J. Chem. Soc. *1954*, 243
110. Kovářová-Lerchová, J., Pospíšil, J.: Eur. Polym. J. *13*, 975 (1977)

111. Chien, J. C. W.: Hydroperoxides in degradation and stabilisation of polymers. In: Degradation and stabilisation of polymers. Geuskens, G. (ed.). London: Applied Science Publishers 1975, pp. 95–112
112. Coppinger, G. M.: J. Amer. Chem. Soc. 79, 2758 (1957)
113. Chien, J. C. W., Vandenberg, E. J., Jabloner, H.: J. Polym. Sci. A–1, 6, 381 (1968)
114. Kovářová-Lerchová, J., Pospíšil, J.: Eur. Polym. J. 14, 463 (1978)
115. Lerchová, J., Pospíšil, J.: Chem. Ind. 1975, 516
116. Lerchová, J., Pospíšil, J.: Angew. Makromol. Chem. 38, 191 (1974)
117. Lerchová, J., Pospíšil, J.: Angew. Makromol. Chem. 39, 107 (1974)
118. Kovářová-Lerchová, J., Pospíšil, J.: Chem. Ind. 1978, 91
119. Dubinskij, V. Z., Roginskij, V. A., Miller, V. B.: Vysokomol. Soed. 18A, 2567 (1976)
120. Roginskij, V. A., Dubinskij, V. Z., Shlyapnikova, I. A., Miller, V. B.: Eur. Polym. J. 13, 1043 (1977)
121. Dubinskij, V. Z., Roginskij, V. A., Miller, V. B.: Dokl. Akad. Nauk SSSR, 220, 1360 (1975)
122. Roginskij, V. A., Plekhanova, L. G., Dubinskij, V. Z., Nikiforov, G. A., Ershov, V. V., Miller, V. B.: Izv. Akad. Nauk SSSR, otd. khim. 1975, 1327
123. Koch, J.: Angew. Makromol. Chem. 20, 35 (1971)
124. Kovářová-Lerchová, J., Pilař, J., Samay, G., Pospíšil, J.: Eur. Polym. J. 14, 601 (1978)
125. Lerchová, J., Kotulak, L., Rotschová, J., Pilař, J., Pospíšil, J.: J. Polym. Sci., Symposia 57, 229 (1976)
126. Pilař, J., Kovářová, J., Pospíšil, J.: unpublished results
127. Dulog, L., Stahlberg, H.: Angew. Makromol. Chem. 74, 285 (1978)
128. Buben, I., Pospíšil, J.: J. Polym. Sci., Symposia 57, 255 (1976)
129. Shlyapnikova, I. A., Dubinskij, V. Z., Roginskij, V. A., Miller, V. B.: Izv. Akad. Nauk SSSR, otd. khim. 1977, 57

130. Buben, I., Pospíšil, J.: unpublished results
131. Campbell, T. N., Coppinger, G. M.: J. Amer. Chem. Soc. 74, 1469 (1952)
132. Ogura, K., Matsuura, T.: Tetrahedron 26, 445 (1970)
133. Ranby, B., Rabek, J. F.: Photodegradation, photo-oxidation and photostabilization of polymers. London: Wiley 1975
134. Lind, H., Winkler, T., Loeliger, H.: J. Polym. Sci., Symposia 57, 225 (1976)
135. Pilař, J., Kovářová, J., Pospíšil, J.: Org. Magnet. Resonance (submitted)
136. Lind, H., Loeliger, H., Winkler, T.: Tetrahedron Lett. 1978, 1571
137. Kovářová, J., Pilař, J., Pospíšil, J.: unpublished results
138. Kotulak, L., Pospíšil, J.: 19th IUPAC Microsymposium "Mechanisms of degradation and stabilization of hydrocarbon polymers", Prague Czechoslovakia, 1979
139. Sedlář, J., Kovářová, J., Pác, J., Pospíšil, J.: unpublished results
140. Virt, J., Rosík, L., Kovářová, J., Pospíšil, J.: unpublished results
141. Lind, H., Loeliger, H.: Tetrahedron Lett. 1976, 2569
142. Tkáč, A., Omelka, L.: J. Polym. Sci., Symposia 40, 119 (1973)
143. Johnston, K. M., Jacobson, R. E., Williams, G. H.: J. Chem. Soc. C. 1969, 1424
144. Ingold, K. U.: Can. J. Chem. 41, 2807 (1963)
145. Bidzilja, V. A., Pokhodenko, V. D., Brodskij, A. I.: Dokl. Akad. Nauk SSSR, 166, 1099 (1966)

146. Cook, C. D., Woodworth, R. C.: J. Amer. Chem. Soc. 75, 6242 (1953)
147. Cook, C. D., Gilmour, N. D.: J. Org. Chem. 25, 1429 (1960)
148. Dimroth, K., Kalk, F., Neubauer, G.: Chem. Ber. 90, 2058 (1957)
149. Nishinaga, A., Watanabe, K., Matsuura, T.: Tetrahedron Lett. 1974, 1291
150. Sklyarova, E. G., Lukovnikov, A. F., Khidekel, M. L., Karpov, V. V.: Izv. Akad. Nauk. SSSR, otd. khim. 1965, 1093

151. Cook, C. D., Woodworth, R. C., Fianu, P. J.: J. Amer. Chem. Soc. 78, 4159 (1956)
152. Müller, E., Ley, K., Schmidhuber, W.: Chem. Ber. 89, 1738 (1956)
153. Trozzolo, A. M., Winslow, F. H.: Macromolecules 1, 98 (1968)
154. Kaplan, M. L., Kelleher, P. G.: Rubber, Chem. Technol. 45, 423 (1972)
155. Belluš, D.: Quenchers of singlet oxygen – a critical review. In: Singlet oxygen. Reactions

with organic compounds and polymers. Ranby, B., Rabek, J. F. (eds). Chichester-New York-Brisbane-Toronto: John Wiley 1978, pp. 61–110

156. Sujakin, A. P., Samsonova, L. V., Shlyapintokh, V. J., Ershov, V. V.: Izv. Akad. Nauk SSSR, otd. khim. *1978*, 55

157. Adams, W. R.: Photosensitized oxygenations. In: Oxidation. Techniques and applications in organic syntheses. Augustin, R. L., Trecker, D. J. (eds). New York: Decker 1971. Vol. II, pp. 65–112

158. Foote, C. S.: Acc. Chem. Res. *1*, 104 (1968)

159. Kearns, D. R.: J. Amer. Chem. Soc. *91*, 6554 (1969)

160. Matsuura, T., Omura, K., Nakashima, R.: Bull. Chem. Soc. Japan *38*, 1358 (1965)

161. Pfoertner, K., Böse, D.: Helv. Chim. Acta *53*, 1553 (1970)

162. Taimr, L., Pospíšil, J.: Angew. Makromol. Chem. *39*, 189 (1974)

163. Foote, C. S., Thomas, M., Ching, T. Y.: J. Photochem. *5*, 172 (1976)

164. Coppinger, G. M.: J. Amer. Chem. Soc. *79*, 501 (1957)

165. Gersmann, H. R., Bickel, A. F.: J. Chem. Soc. *1962*, 2356

166. Gersmann, H. R., Bickel, A. F.: J. Chem. Soc. *1959*, 2711

167. Bickel, A. F., Gersmann, H. R.: Proc. Chem. Soc. *1957*, 231

168. Clough, R. L., Yee, B. G., Foote, C. S.: J. Amer. Chem. Soc. *101*, 683 (1979)

169. Carlsson, D. J., Sproule, D. E., Wiles, D. M.: Macromolecules *5*, 659 (1972)

170. Taimr, L., Pospíšil, J.: J. Polym. Sci., Symposia *57*, 213 (1976)

171. Lichtenthaler, R. G., Ranfelt, F.: J. Chromatogr. *149*, 553 (1978)

172. Taimr, L., Pivcová, H., Pospíšil, J.: Angew. Makromol. Chem. *80*, 149 (1979)

173. Hixson, S. S., Mariano, P. S., Zimmermann, H. E.: Chem. Rev. *73*, 531 (1973)

174. Matsuura, T., Ogura, K.: Tetrahedron *24*, 6167 (1968)

175. Ogura, K., Matsuura, T.: Bull. Chem. Soc. Japan *43*, 3187 (1970)

176. Chien, J. C., Boss, C. R., Jabloner, H., Vandenberg, E. J.: J. Polym. Sci., Polym. Lett. Ed. *10*, 915 (1972)

177. Paris, J. P., Gorsuch, J. D., Hercules, D. M.: Anal. Chem. *36*, 1332 (1964)

178. Matsuura, T., Yoshimura, N., Nishinaga, A., Saito, I.: Tetrahedron *28*, 4933 (1972)

179. Prokofev, A. I., Solodovnikov, S. P., Bogdanov, G. V., Nikiforov, G. A., Ershov, V. V.: Teor. Eksper. Khim. *3*, 416 (1967)

180. Pospíšil, J., Taimr, L., Kotulak, L.: Advan. Chem. Ser. *85*, 169 (1968)

181. Pospíšil, J., Kotulák, L., Taimr, L.: Advan. Chem. Ser. *85*, 191 (1968)

182. Pospíšil, J., Lisá, E., Buben, I.: Eur. Polym. J. *6*, 1347 (1970)

183. Pospíšil, J., Kotulak, L., Taimr., L.: Eur. Polym. J. *7*, 255 (1971)

184. Pospíšil, J.: Chem. Listy *10*, 576 (1958)

185. Ryba, O., Petránek, J., Pospíšil, J.: Collect. Czech. Chem. Commun. *30*, 843 (1965)

186. Ryba, O., Petránek, J., Pospíšil, J.: Collect. Czech. Chem. Commun. *30*, 2157 (1965)

187. Lisá, E., Pospíšil, J.: J. Polym. Sci., Symposia *40*, 209 (1973)

188. Lisá, E., Pospíšil, J.: J. Polym. Sci., Symposia *40*, 233 (1973)

189. Ettel, V., Pospíšil, J.: Collect, Czech. Chem. Commun. *22*, 1618 (1957)

190. Ettel, V., Pospíšil, J.: Collect. Czech. Chem. Commun. *22*, 1624 (1957)

191. Pospíšil, J., Ettel, V.: Collect. Czech. Chem. Commun. *24*, 341 (1959)

192. Pospíšil, J., Ettel, V.: Collect. Czech. Chem. Commun. *24*, 729 (1959)

193. Pospíšil, J.: Sci. Papers, Inst. Chem. Technol., Prague, Organic Technology *4*, Part 2, 491 (1960)

194. Buben, I., Pospíšil, J.: Tetrahedron Lett. *1967*, 5123

195. Buben, I., Pospíšil, J.: Collect. Czech. Chem. Commun. *34*, 1991 (1969)

196. Pilař, J., Buben, I., Pospíšil, J.: Collect. Czech. Chem. Commun. *35*, 489 (1970)

197. Pilař, J., Buben, I., Pospíšil, J.: Tetrahedron Lett. *1968*, 4203

198. Pilař, J., Buben, I., Pospíšil, J.: Collect. Czech. Chem. Commun. *37*, 3599 (1972)

199. Pilař, J.: J. Phys. Chem. *74*, 4029 (1970)

200. Pospíšil, J., Taimr, L., Horák, J.: J. Polym. Sci., Symposia *40*, 319 (1973)

201. Lerchová, J., Pospíšil, J.: J. Polym. Sci., Symposia *40*, 307 (1973)

202. Dulog, L., Vogt, W.: Tetrahedron Lett. *1967*, 1915

203. Lloyd., W. G., Zimmermann, R. G., Dietzler, A. J.: Ind. Eng. Chem., Prod. Res. Devel. 5, 326 (1966)
204. Lloyd, W. G., Zimmermann, R. G.: Ind. Eng. Chem., Prod. Res. Devel. 4, 180 (1965)
205. Foster, R., Foreman, R. L.: Quinone complexes. In: The chemistry of the quinonoid compounds. Patai, S. (ed.), London-New York-Sydney-Toronto: Wiley 1974, pp. 257–334
206. Becker, H.-D.: Quinones as oxidants and dehydrogenating agents. In: The chemistry of the quinonoid compounds. Patai, S. (ed.). London-New York-Sydney-Toronto: Wiley 1974, pp. 335–424
207. Neunhoeffer, O., Heitmann, P.: Chem. Ber. 96, 1027 (1963)
208. Becker, H. D.: J. Org. Chem. 34, 1198 (1969)
209. Horner, L., Weber, K. H.: Chem. Ber. 96, 1568 (1963)
210. Horner, L., Weber, K. H.: Chem. Ber. 100, 2842 (1967)
211. Musso, H., Pietsch, H.: Chem. Ber. 100, 2854 (1967)
212. Sandermann, W., Simatupang, M.: Tetrahedron Lett. 1963, 1269
213. Guryanova, V. V., Kovarskaya, B. M., Miller, V. B. Shlyapnikov, Yu. A.: Izv. Akad. Nauk SSSR, otd. khim. 1971, 289
214. Lisá, E., Kotulak, L., Petránek, J., Pospíšil, J.: Eur. Polym. J. 8, 501 (1972)
215. Dunn, J. R., Waters, W. A., Wickham-Jones, C.: J. Chem. Soc. 1952, 2427
216. Moore, R. F., Waters, W. A.: J. Chem. Soc. 1952, 2432
217. Hartel, H.: Chimia 19, 116 (1965)
218. Yassin, A. A., Rizk, N. A.: Eur. Polym. J. 13, 441 (1977)
219. Shlyapnikova, I. A.: Roginskij, V. A. Miller, V. B.: Izv. Akad. Nauk SSSR, otd. khim. 1978, 2487
220. Allen, N. S., McKellar, J. F., Protopapas, S. A.: J. Appl. Polym. Sci. 22, 1451 (1978)
221. Torsueva, E. S., Shlyapnikov, Yu. A.: Vysokomol. Soed. 19B, 377 (1977)
222. Torsueva, E. S., Shlyapnikov, Yu. A.: Vysokomol. Soed. 20B, 446 (1978)
223. Bruce, J. M.: Photochemistry of quinones. In: The chemistry of the quinonoid compounds. Patai, S. (ed.). London-New York-Sydney-Toronto: Wiley 1974, pp. 465–538
224. Schönberg, A., Mustafa, A.: Chem. Rev. 40, 181 (1947)
225. Bruce, J. M.: Quart. Rev. 21, 405 (1967)
226. Rubin, M. B.: Fortschr. Chem. Forch. 13, 251 (1969)
227. Horspool, W. H.: Chemical Society Specialist Periodical Reports 1, 1 (1970), 2, 1 (1971)
228. Rabek, J. F., Ranby, B.: J. Polym. Sci. 12, 295 (1974)
229. Aleksandrov, A. L., Bubnov, N. N., Lazarev, G. G., Lebedev, J. S., Prokofev, A. I.: Izv. Akad. Nauk SSSR, otd. khim. 1976, 515
230. Ivanov, J. A., Lazarov, G. G., Lebedev, J. S., Serdobov, M. V.: Izv. Akad. Nauk SSSR, otd. khim. 1978, 2134
231. Lazarev, G. G., Lebedev, J. S., Serdobov, M. V.: Izv. Adka. Nauk SSSR, otd. khim. 1978, 2520
232. Tumanskij, B. L., Prokofev, A. I., Bubnov, N. N., Solodovnikov, M. I., Kabatschnik, S. P.: Izv. Akad. Nauk SSSR, otd. khim. 1975, 2816
233. Tsuruya, S., Yonezawa, T.: J. Org. Chem. 39, 2438 (1974)
234. Andrushenko, D. A., Gogdam, L. S., Kachan, A. A., Shrubovich, V. A.: Vysokomol. Soed. 14A, 594 (1972)
235. Kitano, M.: Jap. pat. 71 24, 796 (1971)
236. Rabek, J. F.: Chem. Stosow. 11, 89 (1967)
237. Rabek. J. F.: 23rd IUPAC Congress on Macromolecular Chemistry, Boston USA, 1971
238. Rabek. J. F., Ranby, B.: J. Polym. Sci. A1, 12, 295 (1974)
239. Nakamura, K., Yamada, T., Honda, K.: Chem. Lett. (Japan) 1973, 35
240. Wilkinson, F., Schroeder, J.: J. Chem. Soc., Faraday Trans. II, 1979, 441
241. Orlando, M C., Mark, A., Bose, A. K., Manhas, M. S.: J. Amer. Chem. Soc. 89, 6527 (1967)
242. Petránek, J., Ryba, O., Doskočilová, D.: Collect. Czech. Chem. Commun. 32, 2140 (1967)
243. Foster, T., Elliot, A. D., Adeleke, B. B., Wan, J. K. S.: Can. J. Chem. 56, 869 (1978)
244. Bruce, J. M., Chandhry, A.: J. Chem. Soc., Perkin Trans. I, 1972, 372
245. Sviridov, B. D., Gryzunova, L. P., Kuznec, V. M., Nikiforov, G. A., de Jonge, C. R. H. I., Hageman, C. I., Ershov, V. V.: Izv. Akad. Nauk SSSR, otd, khim. 1978, 2160

246. Cookson, R. C., Frankel, J. J., Hudec, J.: J. Chem. Soc. D, *13*, 16 (1965)
247. Lindsey, A. S.: Polymeric quinones. In: The chemistry of the quinonoid compounds. Patai, S. (ed.). London-New York-Sydney-Toronto: Wiley 1974, pp. 793–856
248. Donet, J.-B., Voet, A.: Carbon black. Physics, chemistry and elastomers reinforcement. New York: Dekker 1976
249. Matreyek, W., Winslow, F. H.: Polym. Prepr. Am. Chem. Soc., Div. Polym. Chem. *16*, No 1, 606 (1975)
250. Hawkins, W. L., Lanza, V. L., Loffler, B. B., Matreyek, W., Winslow, F. H.: J. Appl. Polym. Sci. *1*, 43 (1959)
251. Hawkins, W. L., Hansen, R. H., Matreyek, W., Winslow, F. H.: J. Appl. Polym. Sci. *1*, 37 (1959)
252. Crompton, C. R.: J. Appl. Polym. Sci. *6*, 558 (1962)
253. Vlyugova, M. F., Demidova, V. N., Matveeva, E. L.: Plast. Massy *1978*, No 8, 26
254. Scheele, W., Hillmer, K. H.: Kaut. Gummi Kunstst. *27*, No 2, 43 (1974)
255. Jiráčková, L., Pospíšil, J.: Eur. Polym. J. *9*, 71 (1973)
256. Prusíková, M., Jiráčková, L., Pospíšil, J.: Collect. Czech. Chem. Commun. *37*, 3788 (1972)
257. Jiráčková, L., Pospíšil, J.: Collect, Czech. Chem. Commun. *40*, 2800 (1975)
258. Jiráčková, L., Pospíšil, J.: Eur. Polym. J. *10*, 975 (1974)
259. Jiráčková, L., Jelínková, T., Rotschová, J., Pospíšil, J.: Chem. Ind. *1979*, 384
260. Brunton. G., Gilbert, B. C., Mawly, C. J.: J. Chem. Soc., Perkin Trans. II, 1267 (1976)
261. Bridgewater, A. J., Sexton, M. D.: J. Chem. Soc., Perkin Trans. II, *1978*, 530
262. Farzaliev, V. M., Fernando, W. S. E., Scott, G.: Eur. Polym. J. *14*, 785 (1978)
263. Vodzinskij, Y. V., Vasileva, A. A., Korshunov, I. A., Abakamor, G. A.: Elektrokhimiya 7, 24 (1971)
264. Westfahl, J. C., Carman, G. J., Layer, R. W.: Rubber. Chem. Technol. *46*, 1134 (1973)
265. Stegmann, H. B., Scheffer, K., Müller, E.: Justus Liebigs Ann. Chem. *677*, 59 (1964)
266. Jiráčková, L., Pospíšil, J.: Angew. Makromol. Chem. *70*, 135 (1978)
267. Jiráčková, L., Pospíšil, J.: Angew. Makromol. Chem. *82*, 197 (1979)
268. Tkáč, A., Omelka, L., Kovářová, J., Pospíšil, J.: unpublished results
269. Adams, M., Blois, M. S., Sands, R. H.: J. Chem. Phys. *28*, 774 (1958)
270. Allen, N. S., Bullen, D. J., McKellar, J. F.: J. Mater. Sci. *13*, 2692 (1978)
271. Weinstein, A. H.: Rubber Chem. Technol. *50*, 641 (1977)
272. Jiráčková, L., Pospíšil, J.: Angew. Makromol. Chem. *66*, 95 (1978)
273. Hawkins, W. L., Worthington, M. A.: J. Polym. Sci. *A1*, 3489 (1963)
274. Hawkins, W. L., Sautter, H.: J. Polym. Sci. *A1*, 3499 (1963)
275. Bateman, L., Hagrove, K. R.: Proc. Roy. Soc. *A224*, 389, 399 (1954)
276. Bateman, L., Cunneen, J. I., Ford, J.: J. Chem. Soc. *1956*, 3056
277. Hagrove, K. R.: Proc. Roy. Soc. *A235*, 55 (1956)
278. Kice, J. L.: Sulfur-centered radicals. In: Free radicals. Kochi, J. K. (ed.). New York-London-Sydney-Toronto: Wiley 1973, Vol. II, pp. 711–740
279. Denison, G. H., Condit, P. C.: Ind. Eng. Chem. *36*, 477 (1944); *37*, 1102 (1945); *41*, 994 (1949)
280. Barnard. D., Bateman, L., Cain, M. E., Colclough, T., Cunneen, J. I.: J. Chem. Soc. *1961*, 5337
281. Kice, J. L., Cleveland, J. P.: J. Amer. Chem. Soc. *95*, 109 (1973)
282. Shelton, J. R., Davis, K. E.: Intern. J. Sulphur Chem. *8*, 197, 200 (1973)
283. Ager, I., Barton, D. H. R., Greig, D. G. T., Lucente, G., Samnes, P. G., Taylor, M. V., Hewitt, G. H., Looker, B. E., Mowatt, A., Robson, C. A., Underwood, W. G. E.: J. Chem. Soc., Perkin Trans. I, *1973*, 1187
284. Stendel, R., Lutte, J.: Chem. Ber. *110*, 423 (1977)
285. Kice, J. L., Engelbrecht, R. H., Pawlowski, N. E.: J. Amer. Chem. Soc. *87*, 4131 (1965)
286. Schäfer, K., Bonifačič, M., Bahnemann, D., Asmus, K.-D.: J. Phys. Chem. *82*, 2777 (1978)
287. Kovtun, G. A., Trofimov, G. A., Mashonina, S. N.: Zhur. Prikl. Khim. *49*, 1851 (1976)
288. Aliev, A. S., Farzaliev, V. M., Abdullaeva, F. A., Denisov, E. T.: Neftekhimiya *15*, 890 (1975)

289. Husbands, M. J., Scott, G.: Eur. Polym. J. *15*, 249 (1979)
290. Chien, J. C. W., Boss, C. R.: J. Polym. Sci. *A1, 10*, 1579 (1972)
291. Hawkins, W. L., Sautter, H.: Chem. Ind. *1961*, 1825
292. Armstrong, C., Husbands, M. J., Scott, G.: Eur. Polym. J. *15*, 241 (1979)
293. Scott, G.: Mechanisms of reactions of sulfur compounds, *4*, 99 (1969)
294. Armstrong, C., Plant, M. A., Scott, G.: Eur. Polym. J. *11*, 161 (1975)
295. Scott, G.: Eur. Polym. J., Supplement *1969*, 189
296. Scott, G., Shearn, P. A.: J. Appl. Polym. Sci. *13*, 1329 (1969)
297. Armstrong, C., Scott, G.: J. Chem. Soc. B, *1971*, 1747
298. Scott, G.: Pure Appl. Chem. *30*, 267 (1972)
299. Armstrong, C., Scott, G.: Eur. Polym. J. *11*, 271 (1975)
300. van Tiborg, W. J. M., Smael, P.: Rec. Trav. Chim. *95*, 132 (1976)
301. Koelevijn, P., Berger, H.: Rec. Trav. Chim. *91*, 1275 (1972)
302. Koelevijn, P., Berger, H.: Rec. Trav. Chim. *93*, 63 (1974)

Received August 17, 1979
K. Dušek (editor)

Author Index Volumes 1–36

Polymer Bulletin

Editors:
Prof. H.-J. Cantow, Makromolekulare Chemie, Universität Freiburg, Stefan-Meier-Strasse 31,
D-7800 Freiburg, West-Germany
Prof. J. P. Kennedy, Dept. of Polymer Science, The University of Akron, Akron, OH 44325, USA
Prof. T. Saegusa, Dept. Synthetic Chemistry, Kyoto University, Kyoto, 606, Japan

Editorial Board: H. Batzer, Basel; N. Calderon, Akron, OH; S. Cesca, San Donato Milanese; P. J. Flory, Stanford, CA; J. Furukawa, Tokyo; J. E. McGrath, Blacksburg, VA; H. K. Hall, Jr., Tucson, AZ; H. H. Kausch, Lausanne; T. Kelen, Budapest; M. Kryszewski, Łódź; A. Ledwith, Liverpool; E. Maréchal, Paris-Cedex; J. Meißner, Zürich; A. Nakajima, Kyoto; G. and S. Henrici Olivé, Research Triangle Park, NC; N. A. Plate, Moscow; B. Rånby, Stockholm; C. I. Simionescu, București; S. Sivaram, Gujarat; D. H. Solomon, Melbourne; R. Steiner, Frankfurt/M.; H. Tadokoro, Osaka; M. Takayanagi, Fukuoka; I. Uematsu, Tokyo; C. Wippler, Strasbourg; H. Zahn, Aachen

Editorial Assistant: A. Heinrich, Springer-Verlag Heidelberg

To cope with the rapid progress of polymer science, a new journal is now published characterized by emphasis on rapid publication of papers containing a most concise description of results.
The character of the new journal is between the purely archival journals of full papers and the so-called "letter journals" consisting exclusively of short communications.

Special features:
- rapid publication of papers
- no page charge
- 50 off-prints of each paper supplied free of charge

Subscription information and sample copy upon request

Send your order to your bookseller or directly to:
Springer-Verlag, Journal Promotion Dept.,
P. O. Box 105280, D-6900 Heidelberg, FRG

North America: Springer-Verlag New York Inc.,
Journal Sales Dept.,
44 Hartz Way, Secaucus, NJ 07094, USA

Springer
International

A. Knop, W. Scheib

Chemistry and Application of Phenolic Resins

1979. 111 figures, 88 tables. XIII, 269 pages.
(Polymers/Properties and Applications, Volume 3)
ISBN 3-540-09051-7

The authors present the current theory of phenolic resin chemistry and the technical application of phenolic resins, based day-to-day experience in research, production and marketing, and against the background of economic relevance. Where the first fully synthetic polymers (phenolic resins) stand today and what their future is are subjects of discussion. Looking back at their development, it is shown that after a wide variety of adaptions, they remain technically and economically irreplaceable products with potential for further market growth and a commensurate appreciation of their value. This book will be greatly appreciated by chemists, engineers, marketing professionals, and students.

A. Hebeish, J. T. Guthrie

The Chemistry and Technology of Cellulosic Copolymers

1980. 91 figures, approx. 91 tables. Approx. 500 pages.
(Polymers/Properties and Applications, Volume 4)
ISBN 3-540-10164-0

The driving force behind the great scientific interest in copolymer science and technology is the search for products with useful, new or interesting properties. This monograph provides an informative account of new, improved cellulosic materials and the chemistry and technology involved in their production, as well as the first detailed description of grafted and modified celluloses.
The information contained in this book will be of great value to researchers, manufacturers, but also instructors, interested in the modification of cellulosics for textiles, paper, printing, printing inks, paints, and packaging, as well as in polymerization processes and cellulose derivativization.

A. Gandini, H. Cheradame

Cationic Polymerisation

Initiation Processes with Alkenyl Monomers
1980. 12 figures, 9 tables. Approx. 360 pages.
(Advances in Polymer Science, Volume 34/35)
ISBN 3-540-10049-0

This monograph covers the entire spectrum of initiation systems in the cationic polymerisation of alkenyl monomers. Following a detailed outline of the factors which play an important role in determining the behaviour of cationic polymerisation, each type of initiation is discussed individually. Particular emphasis is placed on the two major modes of initiation: initiation by Brønsted acids and initiation by Lewis acids. The authors analyze the present status of this discipline through a critical review of the literature and a series of specific mechanistic proposals, some of which are entirely new. Published material relevant to the understanding of the processes leading to the formation and characterisation of active species is covered exhaustively. The significance of early work is reinterpreted and the impact of more recent studies as well as their shortcomings assessed. The potentials of new experimental techniques are also discussed. Finally, suggestions are offered for future work in many areas on the basis of the mechanistic proposals developed.
This book will help stimulate further ideas, discussions and research in a discipline which is experiencing a lively renaissance.

Springer-Verlag
Berlin
Heidelberg
New York